Conceptual Revolutions in Science, Vol. 1

A COLLECTION OF SCIENTIFIC EXPLORATIONS
AND INTERVIEWS

by Adam B. Dorfman

Everything in science can be questioned
because it is not a religion.

Conceptual Revolutions in Science,
Vol. 1

by Adam B. Dorfman

Photo Credits: We gratefully acknowledge and thank the following photographers and other sources: Cover image: John Stuart Reid, CymaScope.com. Interior images: CymaScope.com, Creative Commons, SpeakDolphin.com, G. Maxwell, Dr. Long, Kelvinsong, Chris Huh, Chuck Middaugh, Grolltech, and Jeff Dahl. Permission has been obtained wherever the source can be located and has responded to our inquiries. In some instances, there is fair use of copyrighted material. We are always happy to make corrections, include omissions or update information. You may contact us at books@relentlesslycreative.com.

ISBN- 978-1-942790-00-6

Published by Relentlessly Creative Books
Publisher's Website: http://relentlesslycreativebooks.com/
773-831-4944

Dedication

To my Mother, Father and Honey

and all our social media supporters

- people of insight, balance and careful observations.

We are together, to the end of time.

Table of Contents

Acknowledgements

For the development and production of the book itself I feel a deep sense of gratitude:

To my family, for all their love and unique experiences. In particular, I would like to thank my mother. As a former school principal, she taught me at a young age to have empathy for the entire system and its connections. For so many reasons, you have been a great influence in my life.

To the team at Relentlessly Creative Books for seeing the potential and making this book possible.

To Honey Goode, my best friend and research assistant on this book. Throughout the process you have provided a unique voice and made meaningful contributions and I am eternally grateful to have you in my life.

To Daniel Schmidt, Dr. Gerald H. Pollack, Dr. Mae-Wan Ho, John Stuart Reid and MJ Pangman who have provided so many insightful observations for this book. Without your tireless pursuit of the conceptual revolutions and willingness to share observations within the context of my questions, none of this would have been possible.

To our active supporters and followers on our official website, ConceptualRevolutions.com. Your heartfelt desire to pursue a philosophy of broad knowledge and conceptual revolutions in science is what motivated us to publish this book.

We are forever thankful for your support and hope that you will continue to visit our website and connect with us on social media to help us prepare for our follow up book: *Conceptual Revolutions in Science*, Vol. 2.

Preface

While the Ancients made great accounts of their philosophies, it was clear that they were based on a genuine concern for the nature of things and the ability to think broadly. Recent thinkers have instead attempted to subject the phenomenon of nature to specialization. While there may be something gained by a narrow focus, there is also something lost. This book is about recapturing what has been lost.

In this book I have curated scientific discoveries and conducted original interviews with a common goal: to consider the philosophy of broad knowledge and the pursuit of conceptual revolutions.

Our journey begins in the classroom where geometry begins, not with expansive overviews and concepts. In our current educational system, before the learner explores geometry, we require that he or she draw lines. Then, the learner is quickly directed to modern day industrial applications.

Therefore, already at a young age, the learner begins a journey towards specialization having mostly avoided the broad imagery that could awaken him or her to nature's continuous geometric proportions, self-similarities and primordial spirals rooted in and pervasive throughout the entire Universe. From that moment, our knowledge is divorced from that which unfolds to a more connected human experience.

Our system teaches us that right angles, circles and squares are inventions; but they are not. They do not belong to any specific operations; they are discoveries belonging to all life forms.

We must subscribe to the notion that specialization is not a timeless right. It is purely designed to optimize a given mechanical structure until a better way is discovered. None of our structures will ever perfectly describe all natural forces.

Those who preach and practice specialization within traditional structures have often crossed onto a pressured path confined by private interests with the rightful objective of maintaining and optimizing these temporary categories of specialization. While these contributions are critical to the development of exquisite technologies and services, a false narrative of a scientific twilight is created and the humble vast oceans of the unknown is forgotten.

> "What we know is a drop,
> what we don't know is an ocean."
>
> - Isaac Newton

In accumulating these stories, I sincerely hope that what I have submitted here may be read with an open mind and my efforts in addressing these broad subjects may be examined, not with a view to censure, but rather to remedy their defects.

Introduction

I hope the collection of short scientific explorations, interviews and evolving principles presented in this book will provide some light in the search for conceptual revolutions.

We consider philosophy rather than arts and we write not considering man-made but rather natural powers. Therefore, I offer this work in the pursuit of insightful interpretations of data and scientific findings.

In the publication of this book, I sought to interview and feature many scientists who have made scientific discoveries and film makers who have pursued broad knowledge and who in doing so, have demonstrated significant empathy for a more aware, prosperous and connected human existence.

They have not only assisted me in my journey to broad knowledge but have stirred my imagination regarding what is already known and can be pursued.

I think that if you read this book with an open mind, and perhaps an open heart as well, you will discover that many of these fundamental concepts of science are undergoing groundbreaking change and it is exciting to witness and be a part of these conceptual revolutions.

"No problem can be solved from the same level of consciousness that created it."

- Albert Einstein

Chapter 1

Suggestions for Seekers

In your daily life, you may have heard about the principles introduced in this book, but the majority of us have not. Most of us are, like you, curious, really curious, and we want to make meaningful contributions to life and society. You have a story to tell from the observations you make, but in order to tell an eloquent story you must seek your inner world to go beyond what is comfortable.[1]

As Ancient Sanskrit texts have clarified, "behind the thoughts is the one witnessing the thoughts. You are not your thoughts. You must silence your mind to find stillness during the act of observation."[2]

"Only from the heart, can you touch the sky"

- Rumi

If you want to change the world, you need to begin by changing yourself. And to change yourself, you must change the philosophies that guide your broad discussions and reflections.

The principles that we introduce here are designed to help you "be" that change, to observe the world in such a way that it can lead to a multidisciplinary approach to understanding nature.

This book may make you question your ability to observe reality as it is, but in the process, you will grow. And there's always room for growth for us too. We encourage you to send us your own suggestions for future volumes of *Conceptual Revolutions*. For now, what follows is what we suggest.

Leave Room for New Realities

We must teach future generations what we think is true and what is unknown, but we also need to remember that solving problems requires us to leave the door open for future realities.

Understand the Limits of Specialization

Nowadays, many people who provide trusted advice in one field are often completely incompetent to discuss another.

Society has marginalized us to the status of the suits we wear or wealth we acquire, instead of the scope and depth of our understanding and the quality of solutions we provide. Authority is respected for the wrong reasons. Why does it matter whose opinion it is? First, consider the language from start to finish and give merit from a reasonable perspective.

Our ability to doubt and question should not be dismissed. History has proven as much. Seek to acquire broad knowledge and you will be able to follow the evidence, wherever it may lead you.

Beware of the Dishonest Fool

Many of those with vested interests will deliberately ignore or misinterpret broad scientific evidence by removing context. They frequently do this in order to pursue and promote the qualifications of a few who support their interests.

Theories are always updated. So, it is predictable that we'll sometimes discover that some of the stuff we believed was wrong, but was our best guess that moment. An honest fool is all right, but a dishonest fool can be very dangerous. It doesn't matter how convenient your theory is, how loud and intimidating or how smart you are. If it doesn't agree with a valid, repeatable and transparent experiment, it's incorrect. And claiming otherwise, no matter how forcefully it is presented, doesn't make it true.

In your daily conversations, recognize that nothing is a certainty, and everything is a probability or approximation.

Math and Geometry are Discoveries, Not Inventions

Accept math, sacred geometric proportions and the primordial spiral as discoveries, not invention, and you will stumble upon nature's basic principles. If you chose to accept nature as she is, then you'll begin to uncover the secrets of science.

> "Geometry will draw the soul toward truth
> and create the spirit of philosophy"
>
> \- Plato

We maintain that past imaginative geniuses embraced all bodies of knowledge in their development and capacity to solve problems. Clear imagery and language over wide interests of natural wonders' will inspire our future generations.

Seek a Balanced Mind

It is our beliefs that must seek greater balance and cohesion between past, current and future generations to reflect what "is" and not what we would like it to "be."

"Who looks outside, dreams;
who looks inside, awakes."

- Carl Jung

As you silence your mind, you experience the world far from your thoughts by becoming aware of the here and now - the real address of humankind. You will discover that by looking inside you connect to everything on the outside. Your inner world and outer world meet and there will be no more separation. This is where insight is found.

Understand the Meaning of a Word

Know the difference between the meaning of a word and the word itself. Facts and rules utilize the lowest levels of cognitive awareness about the senses. Inform yourself to consider the full context, beyond what is comfortable. Leap over conventional wisdom and reach well-justified conclusions.

Evaluate Yourself Fairly

Look at people in both eyes. Evaluate yourself fairly, not in terms of another's image. Respect each other's free will to pursue scientific questions and create the environment where one can acknowledge one's many honest failures. It is in the act of failing that many scientific discoveries have been made.

Knowledge is Not Power

Knowledge is powerful but it is not power and should not be used to control others: it is simplicity and awareness of a better way forward. We are incomplete sensory beings and we all require broad scientific discoveries and a depth of understanding to find the best responses to daily change.

The Eye of Horus is an ancient Egyptian symbol that represented our six senses, not five: Touch, Taste, Hearing, Sight, Smell and Thought. Watch the "Inner Worlds Outer Worlds" Part 4, documentary with Daniel Schmidt for an expanded explanation on "thoughts."[3,4]

Reward Honesty and Seek the "Likely" Truth, Above All

See those who intentionally withhold critical information about the entire system as humanity's weakness. In your daily activities, do not reward a dishonest character because no system can be sustained without the pillars of trust and integrity.

Everything is connected.

"You are not a drop in the ocean.
You are the entire ocean in a drop."

- Rumi

Nature's Imagination is Far Greater

Nobody knows what this world is about but explore and you will find that nearly everything is interesting, if you go deeply enough. The imagination of nature is far, far greater than the imagination of man.

Beware of the poet who says the beauty of a flower is lost in science. It does no harm to the mystery to know a little more about it. Recognize that the truth is far more marvelous than any artists will ever be able to imagine it to be.

Learn From Your Mistakes, Then Always Look Forward

There is meaning for each individual but it's for you to discover and not for others to choose for you. Learn from your mistakes, keep your eyes open and find your value.

> "You will never be the best copy in the world,
> but you can be the best you."
>
> - Anonymous

Don't Be Easily Fooled; Do the Work Yourself

Never forget to do the work yourself, do not cheat yourself to success and don't be easily fooled. Nothing should be too difficult to comprehend and no one is capable of being much smarter than another.

The science of life is not a complex mathematical formula restricted to a select few. It is a system of elegant, broad and deep connections (by geometric expressions) for "all" to observe and understand.

This section was also influenced by many quotes from Dr. Richard Feynman.[5]

Suggestions on How to Get the Most Out of This Book

To get the most out of your reading, it is important to remind yourself of a critical requirement: if you want to contribute to change, you must first be ready to change yourself. Your contributions to the broad conversations will impact the outside world.

> "The way is not in the sky.
> The way is in the heart."
>
> –Buddha

As you are making a conscious effort to provide meaningful dialogue, remind yourself of these principles. Visualize the possibilities for a more balanced and prosperous world from a collective approach to greater insight for a more resonate system.

At first, read each chapter rapidly to get an overview of the themes and how they interconnect in the flow of diverse topics. This is not a heavy book in terms of weight or pages, but it is dense with information and you may want to use a notebook to write your reflections about the content and any other references to complement your quest. We think this will allow you to internalize your exploration. Once you have completed the whole book, review these sections to remind you of your own thoughts and progression towards a successful journey.

Ensure that you allow yourself to pause, reflect and connect to your personal experiences.

Keep a pencil or pen at your hand to add your thoughts or questions promptly. Circle or highlight a word, phrase or quote you wish to clarify. Marking and annotating your reading provides you the most opportunity to raise your consciousness about a given issue.

If your goal is to access greater insights into the matter of the Universe, you will want to review your notes and findings regularly. Keep the book readily accessible to ensure you can take a moment to review these principles of insight often.

Remind yourself to apply these principles, so they become a habitual way of processing information. This will require a dynamic effort of review and a rigorous approach for its applications.

Apply your new principles at every opportunity to have the greatest impact on the spheres of influence around you. Recognize when a conversation lacks substantial concern and knowledge for the welfare of the entire system. There is no other way.

When observing one of these principles in action, make time to make note of it. As the patterns become more evident, its impact throughout the cosmos will begin to unfold.

Keep a daily or weekly diary of your insightful observations. It will help to consolidate your awareness of life around you, which will in turn deepen your ability to raise your consciousness and individuality. In time you will surprise yourself with a well-documented portfolio of findings.

This systematic self-education approach represents an opportunity for developing one's self-reflection skills and may be an effective tool to help balance the mind.

Nine Things This Book Will Help You Achieve

1. Open your mind to new and stimulating observations

2. Raise your level of consciousness and make new friends

3. Review your goals and ambitions and set new ones

4. Have an increased desire to pursue scientific knowledge

5. Review your inner circles of friends to include more conscientious people

6. Feel more confident and gain people's trust with a more honest approach

7. Cultivate your individuality to become a person of substantial questioning

8. Use self-reflection and meditation as tools to improve your life and impact your surroundings

9. Realize that what you observe matters, you have a voice, you don't need to follow blindly

"For the sense of smell, almost more than any other, has the power to recall memories and it is a pity that we use it so little."

–Rachel Carson

Chapter 2

Beyond Our Traditional Five Senses and the "Mysteries of the Unseen World" Documentary Film

Our path to broad knowledge begins as we explore scientific discoveries that concern our traditional five senses.

It is true the vast majority of the physical world around you, you cannot see, feel, touch, hear or smell given that its energy is either too small, too far, too quick, too slow or invisible. To uncover the unseen physical world, we rely on scientific instruments that are well calibrated to capture moments in time and these new visuals often have a profound impact on the way we perceive nature.

Many are familiar with our traditional five senses – sight, hearing, smell, taste, and touch – but are these the only ways to describe the world, to advance our knowledge and to explore reality? The answer to this question is of profound importance and taking this critical path of investigation is essential to humanity's progress.

It turns out many organisms, including humans, have more than five senses, taking different forms and important qualities. But – before we take you beyond our five senses, let's first investigate the limitations of human sight and hearing and explore a fascinating new discovery about our sense of smell. This will certainly spark your imagination towards a new outlook on the world.

Senses are commonly defined as: "A system that consists of a group of sensory cell types that respond to a specific physical phenomenon, and that correspond to a particular group of regions within the brain where the signals are received and interpreted." Of note, all senses have a related sensory organ.

Sight – The "Visible" Spectrum

Eyes are the organs of vision. They focus and detect images of visible light on photoreceptors in the retina of each eye that then generate electrical nerve impulses to the brain. The typical human eye is only capable of observing light at wavelengths between 390 and 750 nanometers and can distinguish between 2.3 and 7.5 million colors, which is called visible light.[6]

Calling it the "visible" spectrum is a flaw in our scientific language because plenty of animals can see light outside our narrow band. And, relative to the entire light spectrum, the human eye captures only a very small fraction of the total band. As such, your eyesight is very limited and there are many electromagnetic waves of a wavelength that you cannot see.

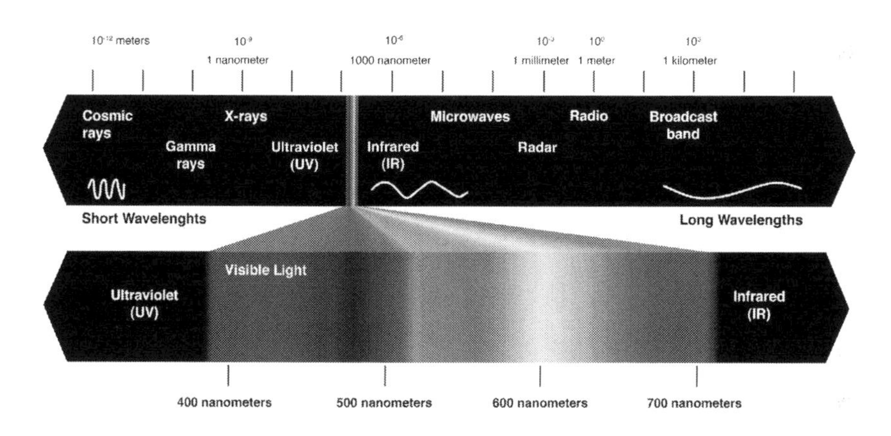

Source unknown

Technologies offer scientific explorers a thrilling new look into long-hidden worlds, allowing viewers to see things not visible to the naked eye. Using high-speed, time-lapse photography and electron microscopy one can peer into the invisible realms of things that are moving too fast, too slow or are simply too small to see.[7]

To discover more about worlds hidden from human sight, check out "Mysteries of the Unseen World" by National Geographic that plays at IMAX theaters worldwide. Louie Schwartzberg, an award-winning cinematographer, who specializes in shooting time-lapse videos, directs the film.

We also encourage you to visit Schwartzberg's personal website at movingart.com. He is always up to something spectacular.

Hearing

The ear is the sensory organ that detects sound. For humans, hearing is normally limited to frequencies between about 20 Hz and 20,000 Hz.[8] However – as compared to many other species, our hearing ability is very limited.

For example, let's consider two animals that we will discuss in this book: elephants and dolphins. We know that elephants can hear the faintest sound and can hear well-below infrasound (20 Hz),[9] while dolphins hear well-above ultrasound (up to 200,000 Hz).[10]

Meanwhile, dolphins are not only richer in higher harmonics but also their range of sonic frequencies. Most human voice sounds range in between about 100 Hz and 10,000 Hz whereas a dolphin's sounds range is anywhere between 100 Hz and 150,000 Hz. That's fifteen times greater than human abilities.

In fact, it's very likely that most of our favorite animals communicate in sound frequencies that are outside our hearing range. We will explore the fascinating possibility of decoding dolphin language in Chapter 4.

Smell – a Major Discovery

The nose is the sensory organ of smell. On March 21, 2014, a team of scientists led by geneticist Leslie Vosshall of Rockefeller University in New York City, systematically tested an almost 100-year-old claim that humans could only discriminate 10,000 odors and instead discovered that our sensory organ of smell could recognize more than 1 trillion odors. That's a big difference and provides a new outlook to understanding the world. These findings not only describe a human olfactory system that outperforms the other senses but it made us wonder how the mainstream scientific community got the previous estimation so wrong.

"We know exactly the range of sound frequencies that people can hear, not because someone made it up, but because it was tested. We didn't just make up the fact that humans can't see infrared or ultraviolet light. Somebody took the time to test it," Vosshall says. "For smell, nobody ever took the time to test."

Per se, the saying that one should stop and smell the roses may in fact have a more literal meaning. The results were published in a report in the journal *Science*, and were conducted on the basis of the results of psychophysical testing.[11,12]

Beyond our Five Senses

Our ability to detect other stimuli beyond those governed by the traditional senses exists, including temperature (thermoception), pain (nociception), balance (equilibrioception), and various other external and internal stimuli.[13]

Additionally, magnetoreception and electroreception are being investigated as human senses. See Chapter 3 for an expanded view on these incredible human abilities.

Conclusion

The purpose of this chapter was to recognize that many outdated views and pre-conceived beliefs about human senses are holding us back. Our existence is not limited to the restrictions of our traditional five senses; new technologies help us develop a greater awareness for nature's entire spectrum and enable us to participate in the Universe in a more meaningful way. The author holds the position that humanity will need to better describe its traditional senses and come to the realization that people have many more sensory abilities whose complementary effects can unlock new worlds of possibilities and understandings.

"We are not going to be able to operate our Spaceship
Earth successfully nor for much longer unless we see it
as a whole spaceship and our fate as common.
It has to be everybody or nobody."

– R. Buckminster Fuller

Chapter 3

Magnetoception and Introducing the *Resonance: Beings of Frequency* Documentary Film

We now set course to explore some fascinating physical and "electrifying" scientific discoveries that extends beyond our traditional five senses.

This chapter is dedicated to discussing new human sensory concepts described in the *Resonance: Beings of Frequency* documentary by James Russell, which can be found on YouTube.[14]

Electromagnetic frequencies are critical to human health, and in fact, new scientific findings have discovered that our cells and

DNA communicate using extremely low electromagnetic frequencies without we could not exist. Furthermore, our brain emits a constant stream of frequencies that we will discuss in Chapter 13.

The premise of the documentary is that the environment is currently being polluted by extremely low frequencies emitted from communication devices, which cause non-thermal negative effects to our health and to the entire ecosystem. It's also known as electro-pollution. To get a better understanding of how this is possible, let's first take a look at some of the most fascinating scientific concepts and natural phenomena discussed in this documentary film.

An extremely low frequency (ELF's) is a subradio frequency. In electromagnetic radiation and health research it is often defined as electromagnetic spectrum frequencies between 0 to100 Hz.[15]

Schumann Resonance – Earth has a Pulse

In 1952, physicist Winfried Otto Schumann discovered that Earth has an electromagnetic pulse.[16] It is now called the Schumann Resonance and it occurs because the space between the surface of the Earth and the conductive ionosphere act as a closed waveguide.

In essence, Schumann Resonance is the principal background in the electromagnetic spectrum and appears as distinct peaks at extremely low frequencies (ELF), around 7.83 Hz.

Scientific evidence demonstrates that this frequency coincides with alpha rhythms produced by the human brain during meditation, relaxation and creativity.[17] Further, scientific evidence strongly suggests that we are healthier beings when we are connected to Earth's natural frequency. How?

This is because when two vibrating systems resonate with each other, like the human brain and the global electric circuit, a rise in amplitude occurs. And, research by Rütger Wever suggests we have greater mental clarity when we tune into Earth's natural frequency of 7.83 Hz.

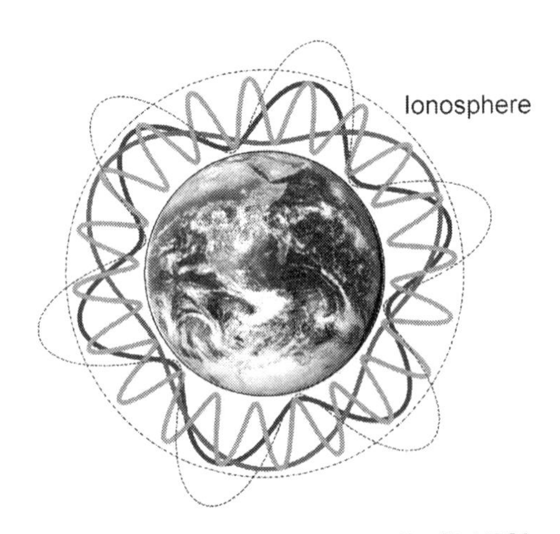

Credit: NASA

Sensitivity to Low Frequency Emissions

In this documentary, director James Russell describes several experiments by Rütger Wever, a German scientist, in support of the view that we are all sensitive to the planet's natural frequency. Between 1964 and 1989, the professor conducted 418 studies in which 447 student volunteers would devote several weeks of time living in a bunker, totally shielded from the natural resonance of the Earth. It was then discovered that without Schumann Resonance the student's physical and mental health suffered. Then, when professor Wever introduced Schumann Resonance of 7.83 Hz, through a magnetic pulse generator, these issues immediately decreased or completely disappeared. As such, Wever's research effectively described a direct connection between human health and the natural frequency of the planet.[18]

> "Measuring Schumann resonance in a city has become impossible. Electromagnetic pollution from cell phones has forced us to make measurements at sea."
>
> –Dr. Wolfgang Ludwig

Regulators and Non-Thermal Effects

According to technology regulators to this day, there is no conceivable mechanism whereby the very-low intensity EMFs could have any biological effects, because the energy involved is below that of the random molecular motions of a system at thermodynamic equilibrium. Not surprisingly, regulators who are responsible for the safety of communications devices seem unwilling to pursue deep and well-funded scientific studies on non-thermal effects to human health.[19]

However, a new science of organism (epigenetics) is emerging that suggests that many non-thermal effects to these very weak electromagnetic fields exist and must be considered. We will discuss the new science of epigenetics in Chapter 14 with leading international scientist Dr. Mae-Wan Ho.

That being said, the electro-pollution consequences may be too late for global honeybee colonies, which are now in major crisis as many seem to have been affected by this pollution.

Bee Colonies in Rapid Decline

Inside the body of bees, a small ring of magnetite particles can be found, which have been proven to detect the magnetic field of the Earth. Bees are hypersensitive to unnatural electromagnetic frequencies (EMF's) and various experiments have shown that artificial frequencies significantly interfere with their biological systems.[20]

By early 2007, abnormally high die-offs (30–70% of hives) of honeybee colonies were reported in North America and in Europe: such a decline is unprecedented in recent history.[21]

While mainstream media and researchers have failed to determine a single definitive cause, this documentary lays the groundwork to suggest electro pollution is causing hypersensitive bees to suddenly die.

This is also bad news for the agricultural industry because of 100 crop species that provide 90% of our global food supply, 71 are pollinated by bees. Should we continue to interfere with the bee's activities, food prices will rise given the need for artificial pollination.

Introduction to Magnetoception

Magnetoception is a sense that allows an organism to detect a magnetic field in order to perceive direction, altitude or location. It has been observed in bacteria, in invertebrates such as fruit flies, lobsters and honeybees. It has also been demonstrated in vertebrates including birds, turtles, sharks and stingrays. Almost all life forms have been shown to have a magnetic sense to varying degrees.[22]

Humans have Magnetosensitive Proteins

Cryptochromes, which are light-sensitive proteins, are found in plants and animals and have recently been found behind the human eye. These proteins are well known to be involved in the circadian rhythms of plants and animals. They are also used to sense the Earth's magnetic fields and pulse.[23]

As a result of this discovery, an experimental study to see whether this human gene could also have magnetic properties was undertaken. Steven Reppert, and his colleagues at the University of Massachusetts in Worcester, removed a fruit flies' usual cryptochrome gene and inserted the human version.

Interestingly, the fruit fly didn't lose its magnetic sense, which confirms the magnetic potential of the human cryptochrome cells. In the journal "Nature Communications," Dr. Reppert later said: "A reassessment of human magnetosensitivity may be in order."[24,25]

The Unseen World has Changed

Much of the research content in this documentary has been verified by a third party peer review process and is controversial primarily because it questions the status-quo of many well-established global industries, not for lack of scientific evidence.

The difference between this documentary and the mainstream view on the research discussed is often in the broad interpretations of the results. Despite these gains and highly relevant findings on electro-sensitive beings, much of this alternative explanation remains hidden from the mainstream population and discourse.

While economic concerns are valid given that electro-frequency pollution has become such a major global industry and has enriched the economic fortunes of many, confronting these issues would only create temporary disruptive forces and allow many healthier and sustainable economic spheres of opportunities to emerge.

Conclusions

The purpose of this chapter was to provide the reader with important scientific findings that describe our connections to nature. The earth pulse is a remarkable phenomenon that clearly affects life functionality and gene expression. This should be taught to all people at a very young age and we believe that incredible industries may emerge with widespread understanding of this simple truth.

The author holds the position that while we can't totally avoid radiation, unnecessary damages are occurring in areas with high concentrations. The telecommunication industry has a significant role in society and mobile technology has taken us to new heights, but requires a better structure interwoven into the ecosystem with a deeper understanding of the new science.

"Vibration underpins all matter in the universe. No matter can exist without sound and vibration."

- John Stuart Reid

Chapter 4

How Dolphins Communicate with Sound Images, Cymatics and "SpeakDolphin" Videos

From our traditional human senses and beyond, we next paddle over friendly waters to explore the thrilling potential of dolphin language and the cymatic properties of sound.

Bioacoustics is a cross-disciplinary science that combines biology and acoustics, and there is an interesting and highly promising research project that is taking place in the United States with dolphins.

This chapter is dedicated to the field of dolphin communication and introduces a new sonic imaging device called the

CymaScope. But first, let's take a look at how sound is described by the mainstream scientific community.

What is Sound?

Sound is a physical phenomenon that cannot exist without a medium, and is often described as a vibration that propagates as a wave, able to travel through all forms of matter: gases, liquids, solids, and plasmas at the speed of sound pertinent to that medium.

As we have previously mentioned, the human experience of sound is limited to frequencies between about 20 Hz and 20,000 Hz.

The True Shape of Sound

Whenever we are shown an illustration of sound it is invariably depicted as a wave, but according to John Stuart Reid, acoustic engineer and inventor of the CymaScope, the term 'sound wave' is misleading and conjures up an incorrect model of sound.

Reid says, "All sounds audible to humans are spherical or bubble-shaped. When you clap your hands, the sound leaves equally in all directions, which infers spheres of sound, not waves. The term 'wave' simply refers to the fact that sound bubbles pulsate in and out in a periodic motion."

Further, Reid maintains that sounds follow the holographic principle such that all points in a sound bubble contain the information needed to describe the entire sound. Here is a visual representation of a sonic bubble enveloping a violin, frozen at a moment in time.

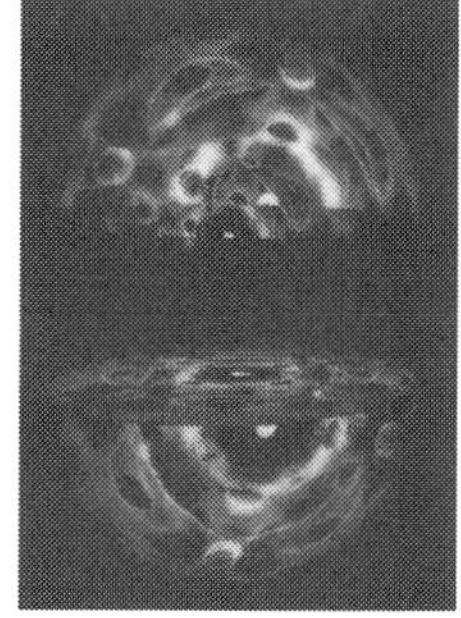

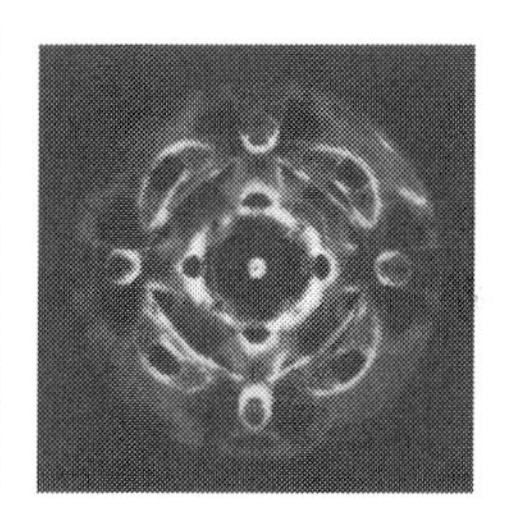

Credit: CymaScope.com

The Cymatics Phenomenon

Cymatics is a generic term for the patterns of vibration that occur on the surface of an object when it is excited by sound.

It is based on the principle that when sound encounters a membrane such as your skin or the surface of water, it imprints an invisible pattern of acoustic energy. When the sound is made visible by one of a number of special techniques, this often creates fundamental and beautiful geometric forms. Put differently, beautiful sounds create beautiful patterns and conversely, harsh or distorted sounds create distorted patterns.

This infers that when we listen to beautiful music we are receiving beauty at a level that is far deeper than our hearing sense alone because according to Reid, the membrane of every cell in our body receives the pattern. How such beautiful or distorted patterns affect us is a question that he hopes to shed light on in a series of microscopic cymatic experiments that are currently underway.

New Technology Emerges – The CymaScope

Using the holographic principles of sounds and cymatics, Reid has developed the CymaScope, the first commercial scientific instrument to make sound visible in a physical medium.

The CymaScope has applications in almost every branch of science simply because sound vibrations underpins all matter. As such, Reid says, "the ability for humanity to see the geometric properties of sound will permit a depth of study previously unavailable to scientists, engineers and researchers."

In Chapter 5, we'll expand on the phenomenon of cymatics with a Q&A with acoustic pioneer, John Stuart Reid.

Water – The Breakthrough

When Reid began developing the CymaScope in 2002, he used fine particulate matter to reveal the cymatic image. Soon after that he discovered that far greater detail and speed of response could be obtained by imprinting sonic vibrations on the ultrasensitive surface of medical-grade water. This is because the surface tension of water has high flexibility and fast responses to imposed vibrations.

Applications for the CymaScope are being identified for many different scientific fields of study, including the sounds of living cells, musical sounds and the sounds of human speech, but it is the language of dolphins where huge strides have been made in recent years.

The Application – Dolphin Language

Whereas most dolphin researchers have been searching for what they believed were words within the dolphin's complex sounds, Jack Kassewitz at SpeakDolphin, instead posed the question: "What if the sounds are not words to be listened to but pictures to be seen?"

The CymaScope team, in collaboration with Jack Kassewitz at Florida-based SpeakDolphin.com, is now working to catalogue the first reliable dolphin picture words. The instrument uses cymatic principles designed to simplify the complex sounds dolphins make into "picture words" that are called CymaGlyphs, each picture representing a dolphin's image for a given object.

Historically, the greatest challenge for marine biologists to capture dolphin sounds has been their sheer complexity. Not only are they rich in higher harmonics but also their range of frequencies is immense in comparison with human sounds. Recall that most human voice sounds range between about 100 Hz and 10,000 Hz whereas dolphin sounds range anywhere between 100 Hz and 150,000 Hz.

Searchlight Beams

Furthermore - while all sounds audible to humans travel in an expanding bubble form and as the frequency of the sound increases above the range of human hearing, the bubbles begin to elongate, resembling searchlight beams.

The team at SpeakDolphin believes that when a dolphin scans an object with its high frequency sound beam, which is emitted in the form of short clicks, each click captures a still image, like a camera taking a photograph.

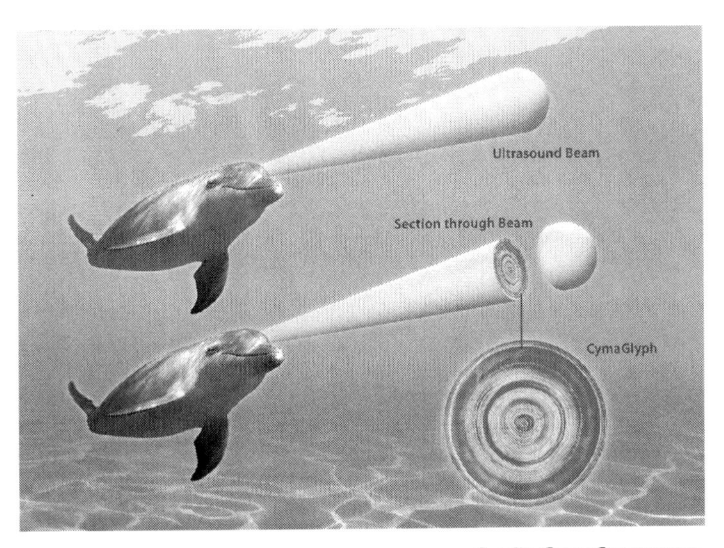

Credit: CymaScope.com

The Experiment that Sparked a New Focus, Intro to SpeakDolphin Videos

In 2011, Jack Kassewitz set up a series of eight submersed plastic objects including a plastic cross, a toy duck and a flowerpot, in a dolphin pool. Using hand signals, he invited one of the research dolphins to 'echolocate' on each object. Kassewitz then recorded the sound as it echoed off each object.

Each sound was then replayed to the dolphin, using a special underwater loudspeaker. Much to his amazement, this dolphin pointed out each object with an accuracy of 86%. Then Kassewitz had the idea to play the same sounds to a different dolphin. He drove two hours to another dolphin facility, set up the plastic test objects in the water and, one by one, replayed each sound to a new dolphin. And, even though it had never experienced these sounds before, it too identified each object with a similar success rate.

Based on the results of the experiment, Kassewitz reasoned that there is a very strong possibility that dolphins may have evolved a language of communication based on sending and receiving sound pictures.

Imaging Echolocation Using the CymaScope

Kassewitz then sent the sound recordings to the CymaScope labs to test the accuracy of this hypothesis. John Stuart Reid began the tricky process of capturing these sounds on the water membrane of the CymaScope instrument. Finally, after several attempts, the instrument was able to pick up the flowerpot image in a rather fuzzy but distinct shape. He went on to image several other dolphin picture words.

While the CymaScope images of dolphin sound pictures are of low definition, Reid is confident that in a short time, with design improvements, the instrument will provide highly defined sonic images that confirm dolphins send and receive. They have named this form of communication "Dolphin sono-pictorial language." Go to their website at CymaScope.com to see the flower pot image.[26]

Leave Room for New Realities

The new science of visual sound – cymatics – will only get better, and requires our support to enable mankind to gain a broader and brighter view on other complex, intelligent non-human life forms on Earth.

Are We Alone?

As Kassewitz said, "Our research has provided an answer to an age-old question, 'Are we alone?' We can now unequivocally answer, 'no.' The search for non-human intelligence in outer space has been found right here on Earth in the graceful form of dolphins."

Conclusions

The purpose of this chapter was to introduce the phenomenon of cymatics; the gateway to making sound visible. Stephen Hawking has said that "physics is about seeing further, better, deeper" and as the age of the CymaScope dawns there seems no doubt that this new instrument will help scientists see better and deeper into the vibrations that underpin all matter.

As previously discussed, our participation on planet Earth is limited by our senses and we require new instrument to describe the unseen world. The CymaScope is one of these instruments that can provide an evolutionary step forward to building a more connected human existence. Incredible wonders will unfold as the cymatics of sounds is pursued and understood.

"If you want to find the secrets of the universe,
think in terms of energy, frequency and vibration."

- Nikola Tesla

"What we have called matter is energy, whose vibration has been so lowered as to be perceptible to the senses."

- Albert Einstein

Chapter 5

Q&A with John Stuart Reid, Inventor of the CymaScope

Our exploration to further describe the cymatics of sound continues here with an interview with acoustics pioneer, John Stuart Reid, inventor of the CymaScope, to discuss ancient knowledge, sound principles and cymatics.

John Stuart Reid is an English acoustics engineer, scientist and inventor. He has studied the world of sound for over 40 years and speaks extensively on his research findings to audiences throughout the United States and the United Kingdom. John's work is inspired by acoustic pioneers, Ernst Chladni, Mary D. Waller and Hans Jenny, and has taken their findings to a new level. His primary interests lie in investigating sound as a formative force and discovering why sound has healing properties.

Cymatics Review

Put simply, cymatics is the name given to the phenomenon that - when sound encounters a membrane, such as your skin or the surface of water, it imprints a pattern of energy. Although invisible to the naked eye the patterns, which are often beautiful, can be made visible with special techniques.

Here is a composite image of 12 piano notes, created by Reid and commissioned by Shannon Novak, a New Zealand-born fine artist as inspiration for a series of 12 musical canvases.

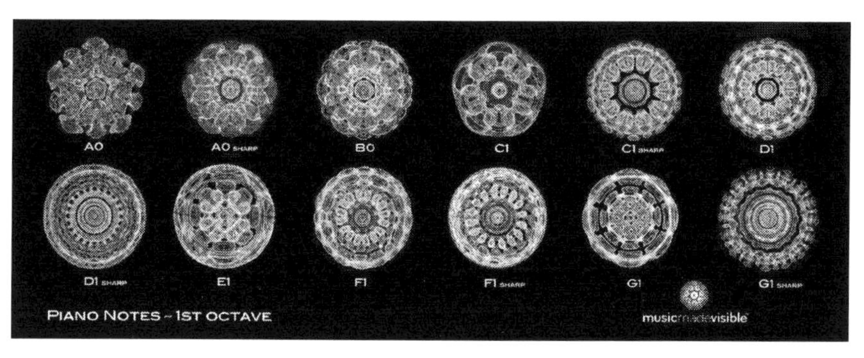

Credit: CymaScope.com

The CymaScope

Using the holographic principles of sound and cymatics, the CymaScope team, led by Reid, has developed the CymaScope, which represents a new breed of scientific instrument that transcribes the periodic vibrations in sounds to periodic wavelets in water, therefore making sound visible.

Entering the Unseen Physical World of Sound

When the microscope and telescope were invented, they opened vistas on realms that were not even suspected to exist. The fields of biology and cosmology would have remained closed to us without these instruments. The CymaScope is the first instrument to give a visual analog of sound (as distinct from a graphic representation given by electronic instruments such as the oscilloscope).

In Reid's opinion, the CymaScope holds the same potential for advancement as the microscope and telescope and its applications are soon likely to touch many aspects of the human endeavor. Visit their website at CymaScope.com for more information.

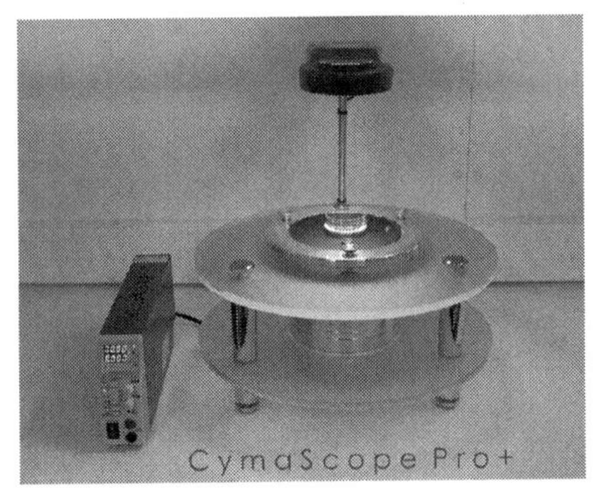

Credit: CymaScope.com

Q&A with Inventor John Stuart Reid

Q1. Is scientific authority respected for the wrong reasons? Have we marginalized scientific authority to a fancy title and language, instead of the quality of solutions provided? Could the scientific community take advantage of this position to pursue self-interests and avoid solutions that challenge the status quo?

A: From my perspective members of the scientific community can be subject to the same flaws as members of all other categories of society, even though the peer review process aims at rigor and fairness, it seems that it is not always achieved. As with all expert and authority figures, we should exercise our best judgment and inner voice before granting our credibility, and give respect where we think respect is due, not as a 'given'. While individual scientists are no more immune to self-interest than anyone else it is common knowledge that giant corporations are sometimes the worst offenders when it comes to discrediting any scientific discovery or direction that does not serve their profits. However, borrowing from Desiderata, "...the world is full of trickery. But let this not blind you to what virtue there is; many persons strive for high ideals..."

Q2. From our viewpoint past imaginative geniuses embraced all bodies of knowledge in their development and capacity to solve problems. What is your feeling towards specialization?

Does a lack of broad knowledge prevent new scientific realities from emerging?

A: Specialization is essential when it comes to developing high technologies simply because the challenges faced in developing such exquisite technologies require it: for example, the piece of high technology that billions of people carry around in a pocket, a mobile phone, or the example of a computer, a flat screen television, or satellite technology that facilitates the internet and worldwide communications. All of these inventions and technologies required teams of specialists to develop the devices to the point where they could be useful to humanity. An individual could never have achieved such results, primarily because most human lifetimes are not long enough to develop all aspects of these high technologies.

However, specialist scientific teams are less likely to put the pieces of giant scientific 'jigsaws' together. Historically, discoveries have been made by individuals who amassed a broad knowledge and were able to see a matrix of observations that led them to put the pieces of a particular puzzle together, to arrive at a ground-breaking invention or discovery. The famous scientist, Peter Plitcha, in his book *God's Secret Formula*, said, "The frontiers of human knowledge have always been pushed back by the individual, not by teams or by heavily-subsidized research programs - a person who can pursue the quest for truth without the pressures and constraints of institutionalized science, and who can furthermore combine expert knowledge from a variety of disciplines."

Q3. Richard Feynman once said, "There's pleasure in discovering old things from a new viewpoint." Do you believe a significant amount of ancient knowledge has been misinterpreted and/or lost?

A: There is growing evidence that ancient peoples had areas of knowledge that are only now being rediscovered by humanity. An example allied to my own field of acoustics is that of the Aboriginals of Australia who have used the sounds of the 'yidaki' (the modern name is didgeridoo) to accelerate the healing of bone fractures and to assist with a range of common illnesses, possibly as far back as 40,000 years ago according to the Aboriginal's oral traditions. While ultrasound (high frequency sound) is a mainstream healing modality that emerged in the 1950's and is used today to support bone fractures and soft injury trauma, the field of audible sound healing is only just beginning to be explored by medical science.

One study, titled "The Felid Purr: A Bio-Mechanical Healing Mechanism," was presented at the 12th International Conference on Low Frequency Noise and Vibration, held in Bristol, UK, in September 2006. The study's focus was the phenomenon of the low frequency purring of injured cats that appeared to speed their healing process. The results of the study were summarized by the statement that vibrations between 20-140 Hz are therapeutic for bone growth/fracture healing, pain relief/swelling reduction, wound healing, muscle growth and repair/tendon repair, mobility of joints and the relief of dyspnea, among other conditions, and confirming that audible sound is, indeed, healing.

Pythagoras also believed in the healing power of sound, in the form of music, and today music therapy is emerging as a mainstream modality, assisting patients physically, mentally, aesthetically and spiritually, and confirming what Pythagoras knew around 2500 years ago.

Summarizing your question, there are many examples of ancient knowledge that have been discounted in the recent past but are now being rediscovered or re-evaluated. I think an important lesson we would be wise to draw upon is not to judge ancient knowledge without first studying it.

Q4. While sounds have been known to have several healing effects, noise pollution can cause serious stresses in a person's life. In fact, changes in the immune system and birth defects have often been attributed to excess noise exposure. How can a better understanding of the geometric properties of sound be used to prevent environmental stress created by noise pollution? See "Noise: A Health Problem," by Dr. William H.

Stewart, former U.S. Surgeon General for an expanded view on noise pollution.[27]

A: Over-exposure to any form of energy is harmful. Sunlight provides a good example; if we bathe in the sun for a few minutes its health-giving benefits are numerous, but overexposure can be very harmful, sometimes leading to carcinomas. The same principle holds true for sound. If we bathe in beautiful music or pure tones at specific frequencies, sound offers many healing benefits, but at high levels, sound and noise can be destructive to human beings.

High levels of ambient noise are not usually sufficient to disorient people, but continued exposure to noise has been shown to cause hearing impairment, hypertension, ischemic heart disease and sleep disturbance. These and other adverse effects could be avoided if society were more aware of the dangers. As with so many risks to health, better education is the key. Most western countries have legislation to govern noise pollution but protection is often limited to employees and does not include non-employees.

The CymaScope can help illustrate these types of risks by graphically making visible the geometry of clean sounds versus the skewed geometry of noisy and distorted sounds. "Beauty begets beauty" in the imagery and conversely, ugly sounds create ugly imagery. But the comparisons go far deeper than the macro realm. Sound affects every cell in our bodies and we are now working to make visible the sound on the surface of living cells.

Q5. One of the most famous musical instrument makers in the world was Stradivarius. These violins and cellos were created in the 17th-18th century and above all, were famous for their quality of sound produced.[28] Many say that no modern-day musical instrument has ever reproduced its sounds. While some research points to wood preservatives used in that day as contributing to the resonant qualities, do you think cymatics could have been used in the instrument's creation? And, how can cymatics impact music going forward?

A: Although there is some evidence that Tibetan gong makers have used Cymatic principles in the tuning of their gongs for over a thousand years and although we know that Leonardo da Vinci and Galileo Galilei were both aware of the phenomenon of cymatics, there is no evidence, as far as I am aware, that musical instrument makers of the Renaissance and later periods used this method of tuning or as a means of improving their instruments. However, cymatics technology is perhaps the most wonderful tool yet invented to support music because it allows us to see a hitherto invisible level of beauty in the music. Without cymatics we would never have known that the sounds of musical instruments contained such beauty and geometric perfection.

In the future I have no doubt that the concept of visible sound will be used by musical instrument manufacturers to hone the designs of their instruments. For example, we are working with a piano company who is using CymaScope imagery to improve each note's sustain (how long a note lingers) and other attributes of their pianos. And cymatics is set to help profoundly deaf people, and other challenged people, in their appreciation of music by means of the CymaScope app for iPhone and iPad, which we currently have in development.

Q6. Scientists at Argonne National Laboratory have discovered a way to use sound to levitate individual droplets of solutions. How do you explain this phenomenon to the layman observer?[29]

A: All audible sounds, including those created vocally by people, propagate in a bubble-like form (not as a wave-like form as is popularly thought). As the frequency of sound goes beyond the upper range of human hearing, say to the high frequencies used by bats or dolphins, the bubble becomes increasingly flattened until it resembles a searchlight beam in its shape. At even higher frequencies (classified as 'ultrasound') sound begins to resemble a laser beam in its shape.

Credit: CymaScope.com

Within these extremely narrow beams of sound there are regions of low-pressure air, known as 'nodes', and regions of high-pressure air, termed 'antinodes'. High-pressure sounds can exert force on small objects and if a matrix of such beams is arranged, an object can be moved up, down, and laterally to any position within the matrix area. (Bear in mind that the objects that can be moved in this way are very small, such as a pea or a drop of water.) The other experiment that is commonly shown in online science videos is a vertically oriented ultrasound beam. A small object is placed in a nodal position in the beam, causing it to levitate in the low-pressure region between two of the high-pressure antinodes.

When the laser was invented someone said, "It's wonderful, but what use is it?" But today lasers are in common use and it is hard to imagine how we ever did without them. I suspect that people are asking the same question now about ultrasound and its ability to levitate and position small objects: "It's wonderful, but what use is it?" All we need is a little imagination!

Q7. One of the most famous sites in the world, Stonehenge, is the remains of a ring of standing stones set within earthworks. Recently, a team of researchers from London's Royal College of Art (RCA), discovered the stones used to construct Stonehenge hold musical properties and when struck, sound like bells, drums and gongs. What is your view on these findings?[30]

A: I have little doubt that ancient peoples were fascinated by some of sound's strange properties. If anyone today makes sound in a cave and hears a voice echo they would think nothing of it because we know that sound bounces off hard surfaces and reflects many times, creating 'mirror images' of our words. But people in ancient times almost certainly thought of the echo as a magic effect or perhaps as linking them with the spirit world. When ancient people made stone burial chambers such enclosed spaces (naturally) also exhibited reverberative qualities, and again we can only imagine how the ancient builders interpreted these acoustic effects. (In prehistory there was no written language, so we have no direct way to know their thoughts on this subject.)

A study of ancient burial mounds of Britain and Ireland by Robert Jahn of Princeton University concluded, in essence, that many of the chambers tested had resonant properties around 110 Hertz, (which is easily sounded by a male vocalist). He was persuaded of the possibility that the ancient peoples may have chanted in the chambers and therefore that the reverberative qualities of the spaces may have been specifically designed, rather than the result of happy accidents. All of this supports the notion that the builders of Stonehenge may also have been aware of the resonant properties of stone.

A team from Reading University in the UK performed a series of acoustic measurements in Stonehenge that showed that sounds made in the interior of the monument were reflected back into the central area due to (they assumed) the deliberate concavity on the inside surfaces of the giant stones. In other words, the builders contrived to enhance the monument's acoustic properties by creating a system of giant acoustic reflectors.

Regarding the team from London's Royal College of Art, concerning the ringing of stones when struck, this feature only works for 'small' stones (of a few tons or less) and even then, only for stones that are largely isolated/decoupled from the ground by their physical attitude, for example, if they happen to be laying on a sharp edge so that most of the stone is free to vibrate. Standing stones, on the other hand, are not able to ring like a bell because being 'plugged-in' to the earth hugely damps them. But archaeology has shown that many ancient people, throughout pre-history, made use of isolated stones, or isolated them deliberately to create sounds; although in most cases we have no way of knowing for sure what beliefs they held regarding the sounds they created in this way.

My own acoustics experiments in the Great Pyramid, published in a booklet titled Egyptian Sonics,[31] led me to conclude that the ancient Egyptians had more than a rudimentary knowledge of acoustics. I was privileged to have been permitted to conduct the study and the results could be the subject of another interview with conceptualrevolutions.com at some point.

Q8. The CymaScope instrument is underpinned by the holographic principle that maintains all points in sound bubbles contain the information needed to describe the entire sound bubble. If all points in the sound bubble contain the same information, how do they differ from each other?

A: In terms of the vibrational information that each air particle (whether an atom or molecule) in the sound bubble holds, there is no difference at all between them. The only aspect of the air particles that differs from one to the next is in their trajectory. A given sound bubble contains trillions of atoms and molecules and each particle carries exactly the same vibrational data but the direction of travel for each atom and molecule is different, naturally, because the leading edge of the bubble requires it to be multidirectional to form the 3D expansion.

In the CymaScope we use the surface of pure water as a membrane onto which a sound is imprinted. The long wavelengths of sound manifest on the water's surface as short wavelets, in a sense the long sound vibrations have been condensed and transcribed to water wavelets, which we can think of as a cross section through the sound bubble. I call these cross-sections “CymaGlyphs” to denote a sound pattern made visible.

Q9. Your team recently made a fascinating breakthrough in the field of dolphin language research. As such, is the CymaScope team working with any other projects in Zoology? And, if so, which project most intrigues you?

A: Our research into dolphin language, which is in collaboration with Jack Kassewitz of SpeakDolphin.com, remains our primary Oceanography/Zoological research project. The dream of holding a conversation with another sentient species is now a big step closer to reality, thanks to the CymaScope instrument. When that day comes, we may be surprised (and perhaps not pleasantly) by what they might have to say to us, given humanity’s history of ill treatment of their species. However, I am happy to hear that many strides are now being made to improve the way dolphins and whales are protected, so if such interspecies communication does become possible we can (hopefully) anticipate having positive forms of ‘communication’ with them.

Credit: SpeakDolphin.com

We have experimentally imaged a range of other animal and bird sounds and there is no doubt in my mind that the CymaScope holds the potential to open up new fields of study in animal and bird communications, although in-depth studies will need to be accomplished by other researchers since our main focus with animal communications will remain with the dolphin, at least for the foreseeable future.

A dolphin in New Mexico recently began mimicking human sounds, as if attempting to speak, and we have imaged some of those sounds, recorded by Jack Kassewitz, which resemble a series of human vowel sounds.

Conclusions

The purpose of this chapter was to expand the readers understanding of cymatics and suggests that many ancient cultures were likely awaken to its physical characteristics. The author's position is that our participation in this Universe requires us to have a greater awareness of sound's resonate frequencies that manifest in all living being. Exploring scientific truth about sound's sacred geometric proportions through cymatics – without respect for old and outdated theories – will not only produce vibrant new industries, but will help restore reality in ways no artist can imagine.

"Discovery consists of seeing what everybody has seen, and thinking what nobody has thought."

- Albert Szent-Gyorgyi

Chapter 6

Q&A with MJ Pangman – Author of "Dancing with Water" and an Introduction to Structured Water

As we resurface from our stunning deep dive, take a moment to take a breath and reconnect with the visible world. Hold your position at the surface and notice a phenomenon called surface tension. The interface between the water and air seems more compact than the rest, because it is.

Up until now – no valid scientific explanation had been given for this observation. In this chapter, author MJ Pangman, joins us to discuss the main observations that better explain this phenomenon. This chapter and interview are dedicated to the conversation of the new science of water and the recent scientific discovery of structured water that may support one of the greatest conceptual revolutions of our time.

MJ Pangman, M.S. is one of the authors of the book *Dancing with Water: The New Science of Water.*[32] Her book is a well-referenced presentation of the new science of water and a guide to naturally structuring, enhancing and revitalizing water for your own consumption. As such, she has done extensive research on the subject and is well positioned to introduce to us the basic findings behind the discovery of structured water.

The Discovery of Structured Water

In 2003, Dr. Gerald H. Pollack and Zheng Jian-Ming, from the Pollack laboratory at the University of Washington, reported an amazing experimental result: They discovered that when you introduced a hydrophilic (water loving) surface next to bulk water a special exclusion zone (EZ) is created - and within this zone, structured water emerged.

The EZ area is so-called because it completely excludes all solutes - substances that dissolved in the water. EZ water has been since named structured water because of its molecular organization.[33]

Since making his findings public, Dr. Gerald H. Pollack, Prof. of bio-engineering, has received the highest honor that the University of Washington at Seattle in the United States could confer on its own staff and awarded a 2008 Annual Faculty Lecture on his research, entitled, "Water, energy and life: Fresh views from the water's edge."[34]

Surface Tension is a Characteristic of Structured Water

Here's an example to help you visualize structured water properties. Take two identical beakers almost filled with water and place them next to each other. Insert an electrode into each beaker to create a high voltage between them. You will notice a water bridge forms and when the beakers are then moved apart slowly, the water bridge stretches and lengthens, but remains intact, even when the beakers are separated by a gap of more than one inch.

What makes the water stiffen up to make a bridge? This phenomenon occurs because surface tension is not bulk water. It's H3O2.[35]

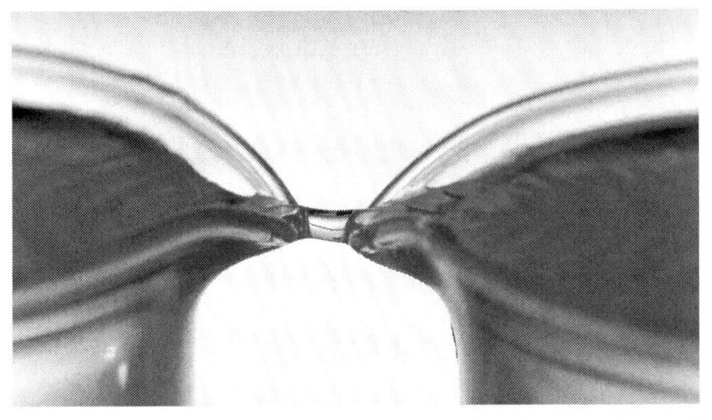

Credit: GNU Free Documentation License 1.2, Gmaxwell

Even though this phenomenon has been well documented, many in the scientific community have mostly ignored this meaningful and new understanding.

H3O2

At the most basic level, bulk water is a substance with a chemical formula of H2O: a molecule with one oxygen atom and two hydrogen atoms, bonded together by shared electrons. However, Dr. Pollack is suggesting that structured water does not have a chemical formula of H2O and is actually H3O2, which makes it more compact, viscous, light absorbing and negatively charged.[36]

As described in Charles Eisenstein's review of structured water – "Pollack and his collaborators hypothesized that the exclusion zone is composed of a liquid crystalline form of water, consisting of stacked hexagonal layers with oxygen and hydrogen in a 2:3 ratio. Of course, ice also consists of stacked hexagonal sheets, but in the case of ice the sheets are held together by the extra protons. Pollack proposes that EZ sheets are "out of register" – aligned so that the oxygens of each layer are frequently next to the hydrogens of the adjacent layers. The alignment is not perfect, but it creates more attractions than repulsions, enough to create cohesion as well as a molecular matrix tight enough to exclude even the tiniest of solutes."[37]

Soon after this discovery, it was also confirmed that the water substance within all living cells is composed of this structured water and its accompanying negative charge seems to be critical for proper cell function. To highlight this point, Dr. Pollack explained, "Cancer cells, instead of being 80 to 90 Millivolts (mV), come in at only around 30 mV."[38]

Structured Water Within the Organism

According to Dr. Mae-Wan Ho, a leading international scientist who confirmed the presence of structured water within organisms, says: "The importance of water in living organisms associated with membranes and other interfaces has been recognized by many pioneers of biochemistry, especially Albert Szent-Györgyi more than 50 years ago. The contribution made in my laboratory was in discovering that organisms and cells are liquid crystalline and coherent to a high degree, even quantum coherent, thanks to the water at the interface. This water is itself structured water and essential to the functioning of biological molecules. Without water macromolecules cannot work at all. Water makes all the difference."

"We don't quite understand interfacial water as yet, but big advances are being made; and bigger advances still, if the conventional community weren't so irrationally afraid to investigate these phenomena. Water, it seems, can be structured in the large as well as in tight places inside cells and tissues.

Water is indeed the basis of a new science. Basic biology and health are poised for a revolution."

Q&A with Author MJ Pangman

Q1: Is authority respected for the wrong reasons? Have we marginalized scientific authority to a fancy title and language, instead of the quality of solutions provided? Could the scientific authority take advantage of this position to pursue self-interests and avoid solutions that challenge the status-quo?

A: In our day, "science" has become the ultimate authority. Articles often start with, "Science reveals," or "scientists say." It gives people a false sense of security in the information that follows. Even more unfortunate is the fact that science (like many areas) has become controlled by money. Funding and publication are the measure of success in science.

Unfortunately, if you want a particular line of scientific inquiry suppressed, you cut funding, you interfere with the publication process, or you discredit the scientist(s). It happens all the time. So it is not really "science" but rather Money that has become the "authority."

Instances of manipulation of the scientific process are all around us. Corporations hire scientists and provide money to academic institutions to fund their research; they place editors on scientific journals to control what does and does not get published; they lobby Congress and spend millions on advertising to convince people to accept their products. Meanwhile, since people are not scientists, they almost need a scientific translator. So corporations send their scientists to deliver the messages they want people to hear. They repeat the message until it carries the authority of "science."

Many of the new ideas that are surfacing today are coming from outside the traditional scientific arena. They have to. So it behooves us to carefully consider new ideas - even if they do not come with peer-reviewed papers or scientific endorsement. More than ever, it is important to realize that the ultimate authority is within each one of us. To claim that authority, is a key to our future.

Q2: From our viewpoint past imaginative geniuses embraced all bodies of knowledge in their development and capacity to solve problems. What is your feeling towards specialization? Does a lack of broad knowledge prevent new scientific realities from emerging?

A: It may not be so much a lack of broad knowledge that prevents scientific breakthroughs, but more the way in which we have traditionally explored the world around us. That's one of the differences between the old and the new sciences. The old sciences spoke to our intellect.

They have limited us to what we could experience with our senses. They have taught us that reality was defined by what we could see, hear and touch - forgetting that our senses themselves are limited. The old sciences have also isolated and compartmentalized every area of science, taking things apart to discover how each part works. Unfortunately, in doing so, the big picture has faded into the background. At the same time, we have lost sight of how each part contributes to the wholeness of the unit, and to the wholeness of life itself.

On the other hand, the new sciences speak to a heart-centered intellect. They can take us to the subatomic world, yet they reveal how all things are parts of the whole. The new sciences show us the inadequacy of our physical senses for evaluating the Universe. They honor both the intuitive and the physical senses - bridging the gap between right- and left-brain as well as the gap between science and spirit. The new sciences reveal that nothing changes without affecting everything else. From this new perspective, we are all participants in the ongoing creative process.

Our feeling is that there will always be a need for specialization, but the key as we move into the future is to keep sight of the big picture and to remember that we are all a part of the greater whole.

Q3: Water and Fluoride – Albert Szent-Gyorgyi said "Life is water, dancing to the tune of macro molecules." Proponents say that water fluoridation helps to prevent tooth decay but what is your view on this? Is it a question of balance?

A: The fluoride ion is found naturally in water to a small degree. But it is found as calcium fluoride, not in the form of fluorosilicic acid, sodium fluoride, or sodium fluorosilicate, which are typically added to municipal water supplies. These forms of fluoride are poisons; they are over 80 times more toxic than naturally occurring calcium fluoride. When fluoride compounds are added to water in the sodium form, it is only a matter of time before the sodium is exchanged for calcium. It rapidly robs the body of calcium. In fact, sodium fluoride poisoning occurs when calcium is stolen from the blood. The anecdote (calcium) provides enough calcium that it does not have to be stolen from the body.

According to a growing number of studies, fluoride is a health risk at any level. When it cannot be immediately excreted, it is taken to parts of the body where it can be sequestered. Since it is attracted to the calcium ion, it goes first to teeth and bones causing dental and skeletal fluorosis. Next it targets the pineal gland as well as nerve and connective tissues.

The fluoride ion is one of the most difficult contaminants to remove from water since it is such a small ion. There are few filtration systems that can adequately remove it without causing other difficulties or environmental problems. We consider the addition of fluoride to be a huge issue where water quality is concerned in the US – especially since many other countries have wisely discontinued its use.

Q4: Alkaline water – It seems like many people have purchased alkaline water treatment devices to neutralize waters acidity. Is this practice prudent?

A: With so much discussion of the need to alkalize, one of the biggest questions people have is whether or not to drink alkaline ionized water. A great deal has been written on the subject by many well-intentioned individuals. However much of the information being circulated regarding the health benefits of alkaline ionized water is either inaccurate or incomplete. First of all, alkaline foods and alkaline water do not neutralize acids in the body. There is no basis for this claim. If they did, then eating overly cooked meat (alkaline) would be the best way to reduce acidity. Cooked meats are alkaline, yet they contribute to over-acidity in the tissues of the body. On the other hand, a lemon (one of the most acidic fruits) alkalizes the tissues of the body. Simply because a substance is alkaline or acidic, does not mean that it will neutralize its opposite in the human body.

pH is all about hydrogen which is the core of life on Earth. The human body extracts energy from hydrogen supplied from organic acids at each stage of the digestive process. Fresh fruits and vegetables are all acidic – with pH 6.9 and under. Digestion releases hydrogen. However, if a person consumes alkaline foods and/or alkaline water, the elimination organs of the body are taxed and free radicals are produced. All the confusion around acidity has led some to believe that acidity is the cause of the problem. In reality, it is part of the solution. Research from Russia shows that when the body needs extra help controlling free radicals, it creates an acidic environment to release hydrogen. What the body is calling for when tissues become too acidic is antioxidants. Hydrogen is one of the most potent antioxidants known.

In Nature, water is frequently found to be slightly acidic. It contains organic acids that are necessary to balance alkaline minerals. The lower the pH, the more minerals water can hold. If the pH rises too high, minerals drop out in the form of scale. Without balance, alkaline minerals wreak havoc in all living organisms. That's why the proponents of alkaline water tell you not to give alkaline water to plants or to put it in a fish tank. It will kill them.

From our perspective, drinking alkaline water is not prudent. It may have some short-term therapeutic benefits, but in the long run it is imbalanced and will contribute to imbalances in the body.

Q5: Dr. Gerald H. Pollack proposes that structured water is H3O2 – As such, does structured water carry more energy to the body's cells?

A: One of the most often reported effects of adding structured water to a person's diet is more energy. That benefit is noted more than all others - especially at first. That could be due to any number of changes at the molecular level as water becomes more structured.

According to Dr. Pollack's research, structured water is electron-rich. Electrons may be considered a source of energy.

Structured water is also more organized. Hydrogen bonds establish a matrix within which energy can be stored and through which information can be delivered. The greater the degree of structure the more efficiently water can store and deliver signals and other information. This efficiency may also translate to more energy.

Structured water is a coherent, fluid crystal where individual molecules cooperate and function as though they were one. Coherence enables each molecule to communicate with every other molecule in the system instantaneously.

Any one of these (or others) could account for greater cellular energy.

Q6: The common basilisk (also called the Jesus Christ lizard) found in Central and South American rainforests is known for its ability to run on the surface of water.[39] How do you explain this phenomenon?

A: In the water molecule, oxygen atoms maintain a slightly negative charge while hydrogen atoms maintain a slightly positive charge. These charges join water molecules together forming an interconnected network. In a large portion of the water on the planet, hydrogen bonding is random and there is no long-term pattern to the way molecules are interconnected. However, it is well known that water molecules at the surface of any body of water are strongly hydrogen bonded. In other words, they stick more tightly to each other than molecular layers below the surface. This produces a slightly stronger surface called surface tension. It is the reason rocks can be skipped on the surface of water; it is the reason a needle can be balanced on the surface of still water; it is the reason some insects can "stride" on the water; and it is the reason the basilisk, moving extremely rapidly, can run on the surface of the water.

Nature provides forces that can influence the degree and the stability of hydrogen bonding beyond the water's surface. Under some circumstances, water develops liquid crystalline structure where organization itself helps to stabilize hydrogen bonds. We refer to this as structured or liquid crystalline water. While water in this phase is still fluid, the enhanced level of hydrogen bonding supports greater cohesive forces with accompanying altered characteristics.

Q7: In the 1969, Dr. Raymond Damadian proposed that cancer cells could be detected with MRI (magnetic resonance imaging) technology. The MRI's success in detecting cancerous cells is given to its ability to measure the mobility of cell water. Structured water is less mobile given its increased organization. As such, what does it mean to the layperson that cancerous cells have less structured water than healthier cells?

A: This offers evidence that there are changes in cellular water accompanying disease. It means that structured water is an integral part of a healthy body. Maintaining its organization (or what we refer to as it liquid crystallinity) is directly linked with health and with youth. Understanding this makes most people feel differently about the energetic quality of the water they drink.

Q8: How much structured water and bulk water is in a glass of water? And, how can someone increase his or her consumption of structured water?

A: All water is structured to some degree. However, there are a variety of factors that can affect the degree to which water is structured. Things like temperature, magnetic and piezoelectric fields, movement (vortexing), light, sound, and even the variety and form of minerals in water, can all affect water's structure and its ability to positively affect a person's body. So it is practically impossible to say that a glass of water has a certain degree of structure without knowing all the factors. But it is not difficult to improve water's structure and its energetic quality. In fact, that's the reason Melanie and I wrote the book, *Dancing with Water*: to let people know how simple it could be to bring all the right factors together (just as Nature does) to return water's liquid crystalline, coherent structure.

Conclusions

The purpose of this chapter was to introduce the new science of water and is an important discussion towards better describing our connection to each other. All living beings contain mainly water, and despite this, we know so little about it. The discovery of structured water and its presence in the organism disrupts the modern view on water and should force us to revaluate its position within our culture. By its very nature, advances in our knowledge of structured water can nourish the roots of our existence while allowing new and more productive industries that better reflect reality to emerge. Now, with these initial discoveries, we should ask ourselves: Has society truly provided the financial resources and runway to adequately pursue the knowledge of this breakthrough discovery? We'll pursue this new dimension in Chapter 15.

"Everything is simply a probability, nothing is certainty."

- Richard Feynman

Chapter 7

An Introduction to Quantum Physics Principles and the 64 Tetrahedron Matrix

Next, we set sails to navigate through basic principles of quantum physics and introduce what appears to be a timeless geometric structure that holds all matter together.

This chapter is dedicated to introducing several quantum mechanical principles and what appears to be a fundamental geometric property shared by all life forms.

Wave-Particle Duality

The double-slit experiment demonstrates that matter and energy can display characteristics of both waves and particles.

This is called the wave-particle duality and is central to current quantum physics principles. Quantum physics is a science of possibilities and approximations rather than exactness of Newtonian physics.[40]

To understand this you must activate your imagination to visualize connecting parts and possibilities within the framework that quantum physics describes.

The Role of the Observer

In the quantum realm the role of the observer is evident by the double-slit experiment. During the experiment, the act of observation by an observer destroys the electrons wave like pattern and disrupts its natural flow of evolution.

This is a major breakdown in the objective reality that believed the observer had no influence or interference with the result. The act of measurement "does something" to the process and influences its evolution.

Whatever that "is" has not yet been explained.

Obstructing Reality

It is critical to understand that an observer influences outcomes as many observers in society are given the task of measuring others. Observers must be careful not to reduce the quantum effect of one's actions to a single linear event and outcome.

Timeless Geometric Structure to Life?

Scattering data collected by the Large Hadron Collider (LHC) discovered a jewel-like geometric object mainstream scientist call "the amplituhedron." Based on this discovery, scientist at the LHC have constructed a mathematical device that drastically simplifies calculations of particle-particle interactions.[41]

The simplification comes from using this geometric structure, instead of doing laborious calculations with thousands of random vector terms to produce a probabilistic outcome.

Consequently, these researchers now suggest that space and time, it would appear, do not provide the fundamental structure of evolution but arise from a timeless geometric structure that confines all matter in this Universe.

It should be noted, the geometric structure described by the LHC, is similar to, but far less detailed compared to the 64 tetrahedron matrix described by Physicist Nassim Haramein, from the Resonance Project foundation. For more information on Nassim's research findings, visit their web site.[42]

Credit: Chuck Middaugh, still from video[43]

Quantum Entanglement

Quantum entanglement demonstrates that there is a physical phenomenon that exists in that pairs (or groups) of particles are generated or interact in ways such that the quantum state of each member instantaneously affects the others. When a measurement is made, the outcome of the other member can be known (i.e. particle spin).[44]

As such, it is now generally accepted within quantum mechanics principles that two particles can impact each other even at large distances and faster than the speed of light.

The Principle of Locality is False

Therefore, the principle of locality, the previous belief that an object can only influence its immediate surroundings, has been proven to be false. As such, given that we cannot visualize all the connecting parts and possibilities, we must act prudently and consider broader interests and understandings when taking action or making decisions.

Quantum Tunneling

Quantum tunneling refers to the quantum mechanical phenomenon where a particle tunnels through a barrier that it classically could not surmount. This plays an essential role in several physical phenomena, such as the nuclear fusion that is believed to occur in main sequence stars like the Sun.[45]

Daily Interactions

As such, with an understanding of the various quantum physics principles, the role of the observer and the recently discovered geometric structure of particle interactions, it seems highly risky for one to reach conclusions solely based on his or her traditional five senses because these senses alone are not enough for anyone to reach well justified conclusions.

Valid, broad observations open to new realities with a consideration of the unique context of another's position should have increased value in the analytic process.

Conclusions

The purpose of this chapter was to convey the importance of openness to new scientific findings that describe various interference patterns in our environment and again introduces the presence of a timeless geometric structure that may hold all matter together. It is the author's position that these discoveries can be used to visualize everyday life. We are all faced with interference patterns that are invisible to others, but it's essential to consider the context of the situation in the evaluation of one's own performance. Evaluate yourself fairly taking into consideration the perspective of these interference patterns, correct your mistakes and then move forward with a balanced mind.

"Most people say that it is the intellect which makes a great scientist. They are wrong: it is character."

- Albert Einstein

Chapter 8

Space Exploration in Our Solar System and Scientific Courage

Our exploration continues as we voyage to space. This chapter is dedicated to the exploration of this solar system, black holes and to the pursuit of scientific courage.

Leave Room for New Realities

Before we begin, let's remember that solving problems must always leave the door open to new realities and the quality of our broad conversation is highly correlated to our ability to solve problems. It should not matter whose opinion it is. We must consider the language from start to finish and give merit from a reasonable perspective.

Introduction to Our Solar System

The Solar System is the Sun and the objects that orbit the Sun. Our planetary system of eight planets and various secondary bodies: dwarf planets and other small objects that orbit the Sun directly, as well as planetary satellites (moons).

The vast majority of the system's mass is the Sun (98%), with most of the remaining mass contained in Jupiter.[46] The two largest planets, Jupiter and Saturn, are composed mostly of hydrogen and helium. The two remotest planets, Uranus and Neptune, are composed of substances with relatively high melting points called ices, such as water, ammonia and methane, and are often referred to separately as "ice giants."

The Solar System also contains regions populated by smaller objects; the asteroid belt, which lies between Mars and Jupiter, and consists of rock and metal. Beyond Neptune's orbit lies the Kuiper belt and a scattered disc composed mostly of ices.

A natural satellite, or moon, is a celestial body that orbits another, which is called its primary. There are 173 known natural satellites orbiting planets in the Solar System: Mercury and Venus have no natural satellites, Earth has one and Mars has two tiny natural satellites, Phobos and Deimos.

The Sun and Earth Travel Through Space

When scientists observe Earth's movements in space, they see that all the planets of our solar system are following the Sun in a heliocentric movement. Consequently, because our sun and galaxy are moving forward through space, the Earth spirals at an incredible distance each year. How far the Earth moves depends on reference points acquired from a space observatory.

As Nassim Haramein from the Resonance Project describes, "Many have been taught the solar system has the Sun in the middle with the planets going around and around in a simple circular orbit. However, not only does that not account for the motion of the Sun around the galactic center, but it also does not include its forward movement."[47,48]

As the earth spirals forward through space, it and the other planets trace beautiful and geometric patterns around the Sun.

Pentagram of Venus

Here is the path of the planet Venus relative to the Earth over an eight-year period. Finding the orbits of any two planets, and then drawing a line between the two planet's positions every few days creates this geometric pattern. Since the inner planet orbits faster than the outer planet, interesting patterns evolve. For your knowledge, eight Earth orbits equal thirteen Venus orbits around the Sun.[49]

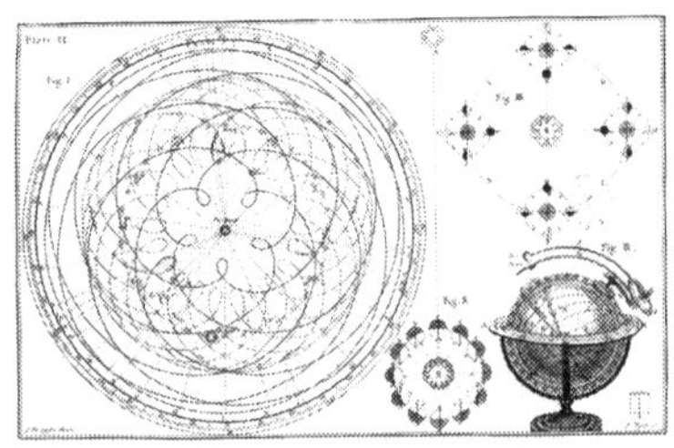

Source unknown

The Black Holes in our Galaxy are not Devouring Monsters

After more than fifty years of theoretical models, society now has a variety of observations of black holes at various scale levels, from stellar to galactic, and cosmological phenomena. When astronomers first began theorizing about black holes, they were expecting that a super massive black hole (SMBH) would consume everything in its vicinity.

Recently however, a team of researchers at UMass Amherst used long observation times (over 5 weeks), and with an improved spectral resolving power from the Chandra instrument (launched in 1999), reached a new conclusion.

According to Daniel Wang from the University of Massachusetts, Amherst, "In principle, super massive black holes suck in everything ... but we found this is not correct. Instead it rejects about 99 percent of this super-hot material, only letting a small amount in, though exactly how it happens is still another question."[50]

More Like Grey Holes

Based on these new observations, Stephen Hawking has also redefined the event horizon of the black hole by saying, "This suggests that black holes should be redefined as metastable bound states of the gravitational field."

In layman terms, Hawking concludes that this effectively means that information can also be emitted from the black hole - making them more like "grey holes," where information (matter and energy) can go in, and come out.

His new conclusion is in direct contradiction to his previous 30-year-old theory that suggested that all the information that fell into a black hole would be forever lost, known as "information loss paradox." As such, we recognize this as a great example of the courage needed to move forward beyond old and outdated theories.[51]

A Misconception: Astronauts Do Not Float in Space

We know there are many challenges and misconceptions about space travel. For one, we often hear that astronauts float in space because it does not have gravity, but this is incorrect. Gravity is a constant in the Universe and is present at all points.

Astronauts at the International Space Station are not floating, but merely falling at high speeds captured in and around Earth's orbit, tracing elliptical patterns in curved spacetime.

As Einstein proposed, spacetime is curved by matter, and free-falling objects in Earth's orbit are moving along straight paths in curved spacetime which are called "geodesics."[52]

Atmosphere of Earth

The atmosphere of Earth is composed of layers of gases confined by the planet's magnetic field. It protects life on Earth by absorbing ultraviolet solar radiation, warming the surface through heat retention (greenhouse effect), and reduces temperature extremes between day and night.

It also becomes thinner and thinner with increasing altitude, with no definite boundary between it and outer space. Hubble and other telescopes orbit in space above Earth's atmosphere to take high-resolution images of very distant stars due to space's near-vacuum state, which eliminates optical interferences.

Credit: NASA

What is the Closest Star?

The nearest star to the Earth, apart from the Sun, is Proxima Centauri, which is estimated to be 39.9 trillion km, or 4.2 light-years away. Travelling at the orbital speed of the Space Shuttle (8km/s or ~30,000 km/h), it would take about 150,000 years to get there.[53] Consequently, we can say that the closest star, apart from the Sun, is very far from us.

Further, stars are not spread uniformly across the Universe, but are normally grouped into galaxies along with interstellar gas and dust. Nebulae are often star-forming regions, such as in the Eagle Nebula.

The Sun and Coronal Mass Ejections (CME)

Coronal mass ejections (CME) from the Sun release huge quantities of matter and electromagnetic radiation into space above the Sun's surface. When ejections are directed towards Earth, the shock wave of the traveling mass can create geomagnetic storms that disrupt the Earth's magnetosphere.

Furthermore, when scientists want to observe how these ejections travel to Earth's atmospheric edges, they can't. It is not well understood how these ejections travel to Earth. Solar winds created by CME's, just appear, after some time delay, in our atmosphere, and magnetically travel through our planet's various field lines. It is believed that "magnetic reconnections" in regions around Earth's magnetic poles then cause spectacles known as the "aurorae."[54]

Simplicity and Awareness to Find a Balanced Future

It is critical for the scientific community to find the courage, within their own spheres of influence, to ensure that interpretations of results are properly anchored in scientific truth rather than old and outdated theories. We should not fear sharing scientific knowledge, but instead fear those who do not share or who have limited the eternal scientific path to a selected few possible answers and investigation methods.

Conclusions

The purpose of this chapter was to help readers develop awareness about the importance of properly anchoring knowledge to the truth while remaining open to new realities to avoid delaying scientific progress and cultural advancement.

It is the author's position that we should not fear sharing scientific knowledge, but instead see the danger in those who do not share or who have limited the scientific path to a selected few possible answers and investigation methods. It is critical for the scientific community to find the courage, within their own spheres of influence, to ensure that interpretations of results are properly anchored in scientific truth (without deference for old and outdated theories) and to follow the scientific truth wherever it may lead.

"The ultimate tragedy is not the oppression and cruelty by the bad people, but the silence over that by the good people."

- Martin Luther King, Jr.

Chapter 9

How Elephants are Non-Human Persons, Altruism and Scientific Culture

Our short voyage to space has concluded, but we're half way through our passage to broad knowledge. We have now landed in the plains of Africa for an intimate exploration with our elephant brothers.

Elephants have been scattered throughout sub-Saharan Africa, South Asia, and Southeast Asia for thousands of years. On these lands, two species are recognized: the African elephant and the Asian elephant. Today, African elephants are listed as vulnerable by the International Union for Conservation of Nature (IUCN), while the Asian elephant is classified as endangered.[55,56]

Today, the World Wildlife Foundation estimates the total population of African elephants is between 470,000-690,000 individuals compared to the Asian elephant population of 32,000. In comparison, a World Wildlife Foundation study suggests 3 to 5 million African elephants inhabited the plains of Africa in the 1930s and 1940s.[57]

In 2012, The New York Times reported a substantial increase in ivory poaching, with about 70% of products sent to China. This increased demand has devastated the African elephant population.[58]

Despite the fact that elephants have displayed advanced culture, consciousness and intelligence to the world's leading scientific minds, our legal systems still regard them as mere objects. They are legal "things" that can be sold for parts. Only humans are afforded basic rights with regards to bodily liberty and integrity and are considered legal "persons".[59]

As such, we now turn our attention to a much-needed conceptual revolution as it regards the legal rights we attribute to elephants. Let's take a look at the scientific knowledge that supports elevating elephants to the status of non-human persons.

Elephants have Culture

Elephants have been known to exhibit a wide range of fascinating behaviors that would describe them as an extremely cooperative, emotionally connected and self-aware person. For example, like humans, elephants must learn behavior as they grow up. They are not born with the instinct of survival.

Not only do parents teach their young how to feed, but also train them to use the tools available to them. The youngsters also learn from their parents their role in the highly complex elephant society. That period of learning will last for about ten years.

Research findings have demonstrated that elephants can recognize themselves in a mirror. This would suggest they possess self-awareness, joining only humans, apes and dolphins as animals with this ability.[60]

Elephants Have Large Brains

With an average mass of just over 5 kg (1 1 lbs.), the brain of an elephant is larger than any other land animal and is very comparable to that of humans in structure and complexity.[61]

While the elephant's cortex has as many neurons as that of a human brain, their cerebrum temporal lobes, which function as storage of memory, are much bigger than those of a human.

Elephants Have Broad Communication Skills

Elephants don't just communicate by touch, sight, smell and sound. Elephants use infrasound and seismic communication over long distances.

In the same way that we will likely SpeakDolphin (from Chapter 4), a future in which we will have basic conversations with elephants could be near, should we not censor or limit the pursuit of evidence.[62]

Elephant Language

The biologist, Andrea Turkalo, has spent the majority of her time over a 20-year period recording the noises that the elephants make using a spectrogram. This device is able to detect lower frequencies that human ears are incapable of hearing and enabling her to record elephant calls previously undetectable. After extensive observations and review of bio-acoustic research, she has been able to recognize the elephants studied by their voices.[63]

Turkalo is currently working in association with Cornell's research on African Forest Elephants. Extensive research in this field is ongoing and one day we will surely unlock the elephant language that exists.[64]

Elephants are Highly Social and Altruistic

Elephants have one of the most closely knit cultures of any living species. Families stay together. They can only be separated when captured or in death. They demonstrate compassion for dying or dead individuals of their kind.

Furthermore, their empathy extends beyond their species as they have been observed to help other species in distress such as humans.[65]

International Bans Increase Ivory's Market Value

Solutions to the problem of poaching focused on trying to better monitor international ivory activities through CITES (Convention on International Trade in Endangered Species of Wild Fauna and Flora).[66]

However, between 1980 and 1990, these legal controls had the very counterproductive effect of raising the product's market value and driving business to black markets.

During this time the African elephant population was reduced from 1.3 million to around 600,000 and the ivory trade was mostly sent to Japan's once booming economy. Throughout that decade, around 75,000 African elephants were killed each year due to the ivory trade, representing worth of around 1 billion dollars in market value.

In 1989, CITES effectively banned the international commercial trade in African elephant ivory by placing the species on Appendix I which is the dedicated list for species that are the most endangered. Once this ban went into effect in 1990, reports indicate the elephant populations in the wild stabilized somewhat. However, the decline in the ivory trade coincides extremely well and is likely mostly attributed with the economic decline of its largest customer, Japan, which at the time was responsible for 40% of global demand.

At the start of the 1990's, Japan's strong economic activity ended abruptly as asset prices collapsed and then hastily declined for 10 consecutive years: it is known as the "Lost Decade." From 1986 to 1991, real estate and stock market prices were greatly inflated due to uncontrolled credit expansion and over confidence. The Japanese economy is still struggling to recover from this period.[67,68]

Today, the rise of China's middle class has again fuelled the market demand for global ivory trade. And, the 1989 international trade ban established by CITES has now further enhanced the market conditions for illegal trade. Some estimate that roughly 90 percent of ivory trade in China is illegal and much of it is now used as medium for money laundering activities. Consequently, the international ban from CITES and other controls have had many adverse effects which includes strengthening the core of exactly that which we wanted to rid ourselves of.

2014 U.S. Government Ban on Ivory Imports

Needless to say, a 2014 U.S. government ban on ivory imports from Zimbabwe is not the broad and long-term solution to the problem.[69]

The reality is that in order to better protect elephants, we require a conceptual revolution that recognizes our involvement in the genocide of non-human persons. Although legislation is required to raise the elephant's status, this discussion is not about that.

This is about inadequate scientific conclusions that have messed up our moral compass. Today, our global knowledge of these animals is far more advanced than it has ever been. With the advent of the Internet and video sharing technologies, the elephant non-human traits and behaviors can be easily explored and described. The global scientific community now needs to take a more courageous stance on these observations to clear the path towards the required conceptual revolution to protect these beings.

Drones?

At the start of 2014, Kenya announced that it was going to launch a new program to address the issue of ivory poachers in its national parks and reserves. Small-fry spy drones were expected to be used in 52 of its national parks and reserves to locate ivory poachers. As reported in the Guardian, this $103 million initiative was to roll out at the end of the same year and came on the heels of a successful pilot program. According to the Kenya government, the surveillance drones reduced illegal ivory poaching by 96 percent by proactively monitoring the movements of both roving poacher gangs and elephants.[70]

While this is an interesting and potentially impactful idea – it does not address the root cause of this problem, which will remain as a culture of scientific ignorance.

Furthermore, as one poacher's market is reduced, eventually another market gains. This is how global markets operate.

We would encourage you to support the non-human rights movement.[71]

Conclusions

The purpose of this chapter was to recognize that we need a conceptual revolution and proper legislation as it regards the status of elephants as non-human "persons." With the arrival of video sharing technologies, the elephant's non-human traits and behaviors can be easily viewed and described by the average viewer and yet still our scientific leaders trail behind.

It is the author's position that the global scientific community now needs to take a more unified and courageous stance with legislators on non-human persons to clear a path towards a more intuitive and connected Universe.

"Fear is not real. It is a product of thoughts you create.
Do not misunderstand me. Danger is very real.
But fear is a choice."

- Will Smith

Chapter 10

How the Great White Shark Can Sense Emotions and an Introduction to WildAid Videos

Next, our journey continues as we propel ourselves beyond the economics of fear toward more fascinating discoveries as it relates to the shark's advanced biological senses.

The most well known shark species, such as the great white shark, tiger shark, blue shark, mako shark, and the hammerhead shark, are considered to be apex predators at the top of their underwater food chain.[72]

As apex predators, sharks play a vital role in our ocean's food chain and yet humanity in recent times has brutally obstructed their ability to play this role. Why? First, let's take a look at scientific discoveries on sharks and Great White Sharks in particular.

Built to Survive Ocean Life

Under a broad definition, it has been recognized that sharks have existed for more than 450 million years. It is estimated that they have since branched out into over 470 species and have been identified as the sister group to the rays.[73]

They can be found in all seas and are common at depths of 2,000 m (7,000 ft.), but are mostly absent below 3,000 m (10,000 ft.).

Sharks are Highly Social, Not Like the Movie "Jaws"

A common assumption is that sharks are solitary hunters who simply travel the oceans in search of food. However, this description of sharks applies to only a few species.[74]

Of particular relevance to this chapter, Great White Sharks can be highly social and remain in large schools from birth. They also display curiosity and play-like behavior in the wild.

Interestingly, sharks struggle to live in captivity and until recently few sharks had ever survived in public aquariums beyond one year. They are often quickly released because they refuse to eat and cannot stand the effects of confinement in public aquariums. Much like an autistic child, the Great White Shark will attempt self-harm by uncontrollably banging its head against the pool walls in attempt to commit suicide. [75,76]

While previous mainstream movies have framed sharks as simple, reckless and violent creatures, the truth is really nothing like that. They are extremely advanced fish with unique capabilities and wisdom.

Ampullae of Lorenzini – Sensitivity to Electric Fields

Sharks have amazing sensors that can pick up electric signals emitted by all living animals in the water. This is also known as passive electrolocation. The Great White Shark can sense frequencies in the range of 25 to 50 Hz and if close enough, it can detect even a faint electrical heart pulse.[77]

The sensor organs responsible for this ability are called the Ampullae of Lorenzini. While Marcello Malpighi first discovered them in 1663, it was 1678 when Stephano Lorenzini provided a more detailed description.[78]

The system consists of hundreds or thousands of pores on the shark's head that are big enough to see with the naked eye. Each canal leads to a small gel-filled chamber "ampullas," which are lined with nerve cells. These cells are likely used as an internal compass, helping them follow the Earth's magnetic field to travel across hundreds or thousands of miles of open water.

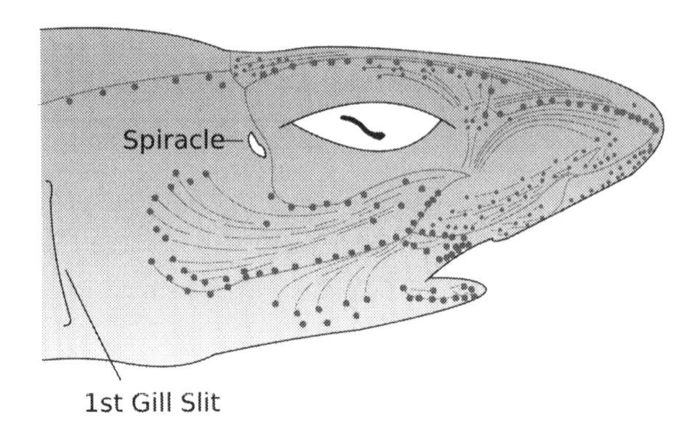

Credit: Chris Huh, Public domain

Can Sharks Sense Human's Brain Wave Activities in Water?

In 1924, German neurologist Hans Berger discovered a way to read brain waves by developing what's known as an electroencephalograph (EEG). And, soon after starting his research, Berger noticed that the electrical activity of each brain wave is correlated to a person's emotional state of mind.

Since sharks have been measured to sense frequencies between 25Hz and 50Hz, which are also related to the emotional status of anxiety and fear, could sharks also feel these brain waves in proximity?[79]

In the Great White Shark 3D movie that plays at IMAX theaters, free-diver William Winram says he believes sharks can pick up on a person's anxiety and fear, and that is what often triggers an aggressive response. Winram observed this sensation while tagging Great Whites for scientific research by holding his breath and free diving with them. You can find many videos on YouTube in which he calmly swims alongside Great Whites, without fear, anxiety nor incident. This would seem to be a very interesting question for future scientist to pursue.[80]

The only thing we have to fear is fear itself.

- Franklin D. Roosevelt

Human Interactions with Sharks

Between 2001 and 2006, the number of human fatalities from unprovoked shark attacks has averaged 4.3 per year globally. Further, it is recognized that most attacks occur during their hunting season. Therefore, it is advisable not to swim in shark-populated areas during hunting season.

Meanwhile, humans are responsible for alarmingly depleting shark populations in our oceans. Based on the analysis of average shark weights, it has been estimated that we are responsible for nearly 100-million shark deaths in 2000, and about 97-million in 2010.

While some species have declined by 90%, it has been estimated that over the past decades the overall population has declined by ~70%. This is both shocking and unsustainable.

Shark Fin Soup – Introducing WildAid's Public Awareness Campaign

At the present time, most sharks are killed for shark fin soup. Shark fins have become a major trade in black markets all over the world. Fishermen capture live sharks, fin them with a hot metal blade and dump the remaining 98% of the animal back into the water. The immobile shark soon dies from suffocation or predators.[81]

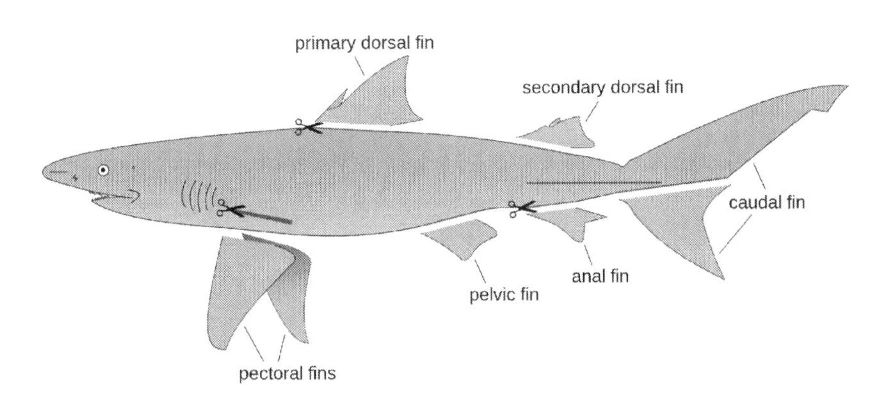

Credit: Grolltech, Creative Commons Attribution-Share Alike 3.0

In many Asian countries, those in the rising middle classes see shark fin soup as a status symbol. As such, a fear of social rejection drives demand in China.

In response to this issue, global icons like Yao Ming are working with conservation groups such as WildAid to engage the Asian consumers in dialogue and to help awaken them to the ignorance of their actions. Check out their website and videos.[82]

The Importance of Facing New Realities and the Limits of Specialization

So why do we obstruct and interfere with sharks? I'll suggest that many leaders, older workers and younger apprentices that work in this shark-finning culture (and all related industries who oppose deterring legislation), are fearful that given their specialized business practices and our highly specialized society, they will find themselves without work and unable to transfer their specialized skills to other industries. Their industry is now an economic bubble set to collapse, which does not provide long-term solutions for the people who work in the industry. There are few paths to regenerating job skills and few economic solutions for the hardships of the many who will be unemployed when there are no more sharks left to fin.

This is the limit of specialization. Authorities must not only legislate this industry but also provide better alternative paths for those involved to acquire more transferable skills.

So, we have a new reality and our past view of sharks has been proven to be completely incorrect. Now we must face the understanding that the fallacy of all our global fears has supported a global shark finning industry that is now completely out of control.

Conclusions

The purpose of this chapter was to provide another great example of the benefits of having a balanced mind. Free divers around the world have few issues swimming with predator sharks so long as they keep their emotions under control. The author believes this situation can be used symbolically in our personal lives to represent the significant danger of finding comfort with one's emotion of fear. The Great White Shark is a vital contributor to our ecosystem with fascinating capabilities that we must learn from, and yet still we pursue so little genuine knowledge. We encourage the reader to consider the broad context and go beyond these fears in the pursuit of the scientific truth to find a better way forward.

"The problem of power is how to achieve its responsible use rather than its irresponsible and indulgent use - of how to get men of power to live for the public rather than off the public."

- John F. Kennedy

"For to be free is not merely to cast off one's chains,
but to live in a way that respects and enhances
the freedom of others."

- Nelson Mandela

Chapter 11

Did Early Athenian Rhetoricians Abuse Democracy to Free Themselves Only to Enslave Others?

Our voyage now takes us back in time to the early moments of Democracy. This chapter is dedicated to discussing how Athenian moral reforms set forth a golden age, only to be brought down by an emerging class of Athenian rhetoricians.

Who are Rhetoricians?

In short, we describe rhetoricians as people who deliberately ignore or misinterpret broad scientific evidence by removing context to pursue and promote self-interests. Over the long-term, these tactics are not repeatable and the consequences may become irreversible. In a free society, dishonest information accumulates to a tipping point that destabilizes the social and economic pillars of trust and integrity towards collapse.

Ancient Philosopher's Description of Rhetoricians

Socrates considered "orators (rhetoricians) and tyrants," as two who belong within the same category.

Aristotle best described rhetoricians as "a counterpart of both logic and politics," and "the faculty of observing in any given case the available means of persuasion."

Plato explained that morality or "Ethics" is not inherent in rhetoric and that Rhetoricians "who would teach anyone who came to him wanting to learn oratory, but without expertise in what's just."

Democracy in Athens

The Athenian democracy is often described as the most significant and well-documented example of first democracy.

An important figure in Athenian democracy is Solon (638 BC-558 BC) known as an Athenian statesman, lawmaker, and poet. In his early years, many Greek city-states had been threatened by the emergence of tyrants, opportunistic noblemen who sought power on behalf of small groups of private interests.[83]

In his early poems, Solon portrayed Athens as being menaced from uncontrolled greed and the arrogance of its citizens. The majority of Athenian's had become slaves to the rich and the feuding of the aristocracy had stalled economic growth and fuelled a cultural decline.

Aristotle, in the Accounts of the Constitution of the Athenians, dictated that debtors unable to repay creditors would be required to surrender lands and give one sixth of produce to their creditors. Should the debt be deemed to exceed the "perceived" value of debtor's total assets, the debtor as well and his family would become the creditor's slaves.[84]

In 594 B.C., Solon was chosen as a temporary archon (ruler) to change the tides and to resolve Athens economic decline. It is during this time that historians believe the first steps to democratic rule were taken, as Solon would permit all Athenian citizens to participate in the assembly, regardless of class.

The Athenian Assembly

In Athen's direct democracy, citizens did not nominate representatives to vote on legislation and executive bills on their behalf (as in the United States), but instead voted as individuals at the assembly. However, "all citizens" were still by no means inclusive of "all" (woman were excluded, for example).

In a move to appease the ruling class, despite granting common voting rights to Athenian men in the assembly, Solon also set up a rival council of 400 men called the Boule who retained power to set the legislative agenda for the main assembly and to protect the property rights qualifications of the few.

Those in the lowest class could not join this group. Furthermore, although, under these new reforms, the position of ruler "was opened" to others, it would also be limited to those who met certain property qualification requirements.

Moral Reforms

Despite these contradictions, Solon effectively re-vitalized an Athenian economy, which overly favored lenders with moral reforms. He ordered widespread debt relief, cancelled all outstanding debts and freed all Athenian people who were victims of serfdom. These actions would eventually drive the Athenian people to then unseen economic and cultural growth before declining again as the old ways returned in later years.

Solon's reforms were later known as the Seisachtheia (shaking off of burdens). Furthermore, accounts suggest these new laws would prohibit the use of personal freedom as collateral in all future debts.[85]

Emergence of Cleisthenes

A better democracy emerged when Cleisthenes, a noble Athenian of the Alcmaeonid family, further reformed the constitution of ancient Athens in 507 BC. In a swift move, he increased the power of the Athenian citizens' assembly and redistributed the power of the nobility over Athenian politics. He did so by changing the political organization and power structures, going from the four traditional tribes under Solon, which were based on family relations, to ten tribes divided according to their area of residence.

Cleisthenes Athenian democracy and a new set of laws were written with the intent to move the nation towards a structure in which no property right qualifications would influence the vote.

Fast forward to our modern democracy; while most western nations provide citizens equal rights to vote on representatives, we allow elected officials to take secret financial contributions from lobbyist groups in exchange for votes on legislation that favor the property qualifications of the few. Thus, the ability to vote with property right qualifications remains.[86]

Cleisthenes' democracy has never been achieved.

The Emergence of Themistocles

Themistocles (c. 524–459 BC) was an Athenian politician and a general. He was a new brand of non-aristocratic politicians who rose to distinction in the early years of Athen's new democracy. As a politician, he had the backing of lower economic class Athenians, and generally was at odds with the Athenian nobility.[87]

Soon after his election as an archon in 493 BC, the Athenian people found new fortune in nearby silver mines that would greatly subsidize Athen's "golden age".

With new financial means, Themistocles received support from the Athenian people to build a fleet of 200 triremes to defend the people from neighboring Greek invaders.

However, it was against the second Persian invasion of Greece that Themistocles led the Athenian navy to its greatest victory establishing Athens as a military power and regional leader.[88]

For this, Themistocles looked for recognition. Accounts suggest his rhetoric began to pursue and promote his short-term interests above concerns of the entire system. Various documents suggest that he would often accept additional bribes on the grounds of special considerations for his past achievements and state title. In the end, we suggest the Athenian lower and upper class, not only permitted it to "be," but "all" would soon begin to emulate this behavior.

Athenians Would Free Themselves, Only to Enslave Others

Following their new military successes, Athens would lead The Delian League, an association of Greek city-states founded in 478 BC. This league allowed military resources from over 150 to 173 member states to be shared. Subsequently, for much of the 5th Century, instead of freeing others and maintaining cohesion, Athenians rhetoricians democratically fed off an empire of subject states.[89]

Solon's empowerment of local laboring classes would also be replaced by an extensive use of imported non-Greeks as chattel slaves.[90]

During this time, there would also be rare limits on the power exercised by the assembly. These democratic rulers acted with dangerous brutality, as in the decision to execute the entire male population of Melos and sell off its women and children for refusing to become subjects of Athens.

Those who opposed or tested the power structures would be ostracized. Ostracism was an extraordinary practice that contrasted with Athenian law at the time: there was no charge, and the individual expelled could present no defense. It was simply a command from the Athenian people by secret vote to banish a person for a ten-year period.

A great example of the emergence of rhetoricians at the time would be from ancient authors who, mostly from elite backgrounds, began to redefine the meaning of democracy so that "demos" in democracy meant not the whole people, but the people who were not the elite. Consequently, the elites were not holding themselves accountable and considered themselves above the democratic structure.

Based on this outlook, groups of oligarchs would frequently battle and emerge knowing that they could ignore the democratic process for personal gains.

The Execution of Socrates

In an attempt to counter the rising disinformation tactics by these influential people, he named "rhetoricians," Socrates developed an alternative method of teaching, known as the Socratic Method. This technique was designed to help restore reasoned arguments, establish context and improve problem-solving abilities.

Its structure is now the supporting pillar of our current "scientific method," but for his efforts, in 399 BC, Socrates was put on trial and executed for "corrupting the young and believing in strange gods." His death gave Europe one of the first intellectual martyrs.

The "Socratic Method" is Today Known as the Scientific Method

The dialectical "Socratic" method was designed for discourse between two or more people that held different points of view on a given subject. Its purpose was to establish the truth guided by reasoned arguments.

Socrates understood the quality and honesty of our broad dialogue ultimately dictates the cohesiveness of society. Our ability to explore the reasoning used in opposing views is highly correlated to our ability to solve critical problems that concern or threaten the sustainability of the entire system.

The Socratic Method was described as: searching for general, commonly held truths that shape opinion and scrutinize them to determine their consistency with opposing beliefs to balance what is most important. It was a form of inquiry and discussion between individuals, based on asking and answering questions to stimulate critical observations and to seek new ideas.[91]

The Fall of The Greek Empire

A little more than 60 years after Socrates was executed for seeking truth guided by reasoned arguments, Macedonia would conquer Athens in 338 BC and dissolve the Athenian government in 322 BC. Greece would then be absorbed into the Roman Republic in 146 BC as part of the Achaea Province, concluding 200 years of Macedonian supremacy.

The Fall of The Roman Empire

By 100 AD, Rome itself adopted rhetoric as an important way of public life. This movement was led by Cicero (106-43 BC) who was one of its most famous ancient orators. Published around the same time, "*Rhetorica ad Herennium*", which is the oldest surviving Latin book on rhetoric, is still used today as a reference textbook on the structure and usage of rhetoric and influence in Law.

On August 24, 410, nearly 300 years after rhetoricians achieved popular acceptance, Rome was sacked for the first time in over 800 years by the Visigoths, led by Alaric I.

We conclude the fall of these two great empires were likely the results of disinformation tactics from those who exercised their property right qualifications within these empires' many spheres of influence. We believe, that in the end – it was an accumulation of poor conversations and reflections that lacked empathy for the entire system that weakened and ultimately destroyed these once glorious political and cultural structures.

Rhetoricians: Belong to the Theater, Improvisation & Comedy Houses

Rhetoric is oratory art that requires neither logic nor proof and is a destructive method to argue one's position for all matters, especially those that concern the entire system. It belongs to the theater, improvisation and comedy houses, not in the leadership nor decision-making process.

Why are comedians so apt to see through the dishonest rhetoric from influencers? Because, they better understand how to pretend to have character and pretend to solve problems. It's an inherent part of their craft.[92]

Conclusions

The purpose of this chapter was to introduce the concept of rhetoricians who were described by Socrates and emerged from the early Athenian democratic movement. We describe these individuals as people who deliberately ignore or misinterpret broad scientific evidence by removing context to pursue and promote self-interests. We suggest that it's the poor quality of the broad conversation due to the dominating voice of rhetoricians that led to the fall of the Roman and Greek Empires because no system can be sustained without trust and integrity.

It is the author's position that one should always beware of rhetoricians and as such, our ability to doubt and question authority should never be questioned. History has proven as much. We encourage the reader to pursue broad scientific knowledge and higher-quality conversations that recognize the full spectrum of the environment with the absolute purposes to follow the evidence, wherever it may lead. We must learn from and recognize our mistakes and then move forward.

"The best and most beautiful things in the world cannot be seen or even touched. They must be felt with the heart"

- Helen Keller

Chapter 12

Telepathy, The Institute of HeartMath and New Brainwave Technologies

Weaving through time, we now travel to Ancient Egypt to explore the meaning of the Eye of Horus. This chapter is dedicated to the discussion of brainwaves, the heart's electromagnetic field, the pineal gland as a sensory organ and how they may all relate to this Ancient Egyptian symbol.

Brainwaves

Emotions are the communication links between neurons within our brain that give birth to all of our thoughts. When masses of neurons communicate with each other through a synchronization of electrical pulses, they create brainwaves.

Today, we are able to measure such activity by using an instrument called an electroencephalogram (also known as an EEG).

Most EEG devices detect brainwaves by placing sensors on the scalp. The information acquired from these measurements is subsequently divided into bandwidths that can describe a function. Perhaps the best way to imagine brainwaves is as a continuous spectrum of consciousness.

Research has found that not only are brainwaves representative of our emotional state, but they could be simulated to help change a person's mental condition, which in turn can help with a variety of mental issues.

Discovered in 1875

Although they were found in 1875, the science of brainwave activity remained a very mysterious subject of study for many more years. In 1924, German neurologist Hans Berger discovered a way to read these electrical currents by developing the electroencephalogram (EEG). Soon after, he realized brainwaves were highly correlated to a person's emotions.

Why did our brains evolve to transmit discrete quantities of electromagnetic frequencies that can be measured to recognize our emotional condition? Many experimental studies support a very practical role for neural oscillations, but a unified interpretation is still lacking.

One of the first discovered frequency bands was alpha activity (8-15 Hz). Afterwards, other frequency bands were found such as: delta (1-3 Hz), theta (4-7 Hz), beta (16-31 Hz) and other faster rhythms such as gamma (30-100 Hz) activity that has been linked to cognitive processing.[93]

New Products Emerge Powered by Brainwaves

Many products and services are emerging to help individuals benefit from emotive brain waves. These commercial EEG devices found on the market are equipped with tiny sensors that sit on the surface of the brain and capture these electrical signals to facilitate man to machine instructions.[94]

It's Not Just the Brain that Emits Electromagnetic Radiation

According to Dr. Rollin McCraty, Director of Research at the HeartMath Research Center, the heart's magnetic component is about 100 times stronger than the brain's magnetic field and can be detected several feet away from the body using a Superconducting Quantum Interference Device (SQUID)-based magnetometers.

Meanwhile the electrical field, which is measured by an electrocardiogram (ECG), is about 60 times greater in strength than the brain waves recorded in an electroencephalogram (EEG).

Furthermore, their research explored interactions that take place between one person's heart and another's brain and discovered that one person's brain waves can synchronize to another person's heart.[95]

Essentially, their findings suggest there is some electromagnetic information being exchanged between people that is influenced by our emotions. Check out the work at heartmath.com and see their YouTube page for more Heart Science videos.[96,97]

Simulated Emotions to Improve Mental Health

Of particular interest to us are new programs from the Institute of HeartMath (IHM), which not only leverage information provided by EEG and ECG devices to simulate emotions, but help people maintain emotional balance and happiness. Not surprisingly, at the moment these instruments are frequently being used to help military veterans rediscover emotional coherent reference points and to recover from Post-Traumatic Stress Disorder (PTSD). We feel these applications will expand to help treat people in all walks of life who struggle with emotional and stress disorders.[98]

"Our scientific power has outrun our spiritual power. We have guided missiles and misguided men."

- Martin Luther King Jr.

Is Telepathy Possible?

What is clear is that the brain and heart emit a tremendous amount of extremely low electromagnetic frequencies that are highly correlated to our emotional and functional condition. Can humans really receive and interpret another's emotional electromagnetic information?

For this to occur, we would need to identify a sensory organ that is able to capture these signals. New evidence suggests the pineal gland could be this sensory organ. In 2004, Dr. David Klein, Chief of the Section on Neuroendocrinology at the NIH, discovered that the interior of the pineal gland has retinal tissue composed of rods and cones (photoreceptors) lining just like the eye, and is even wired into the visual cortex of the brain. He also noted that it even has vitreous fluid in it.

Within this context, it would seem that the scientific community may need more courage to follow this evidence wherever it may lead in the pursuit of scientific truth, especially where it concerns the pineal gland's role as a human sensory organ, and to forego old and outdated theories.[99]

"The photoreceptors of the retina strongly resemble the cells of the pineal gland."

- Dr. David Klein, Science Daily

Thoughts are Physical

Traditionally, people cannot accept emotive telepathy because they do not believe thoughts have a physical representation. As such, the traditional definition of telepathy has been "by the process that transmits information through 'thoughts' from person-to-person without the use of our traditional five senses or physical interaction."

But, as we have introduced, thoughts *are* physical.

The Difference is in the Markup Mechanism

The confusion lies in the markup mechanics of these electromagnetic forces. As these new devices suggest, we do not hear thoughts, but rather feel them through the physical transfer of quantities of electromagnetic information.

The Eye of Horus

The Eye of Horus is an ancient Egyptian symbol that represented our six senses, not five: touch, taste, hearing, sight, smell and thought. Watch the "Inner Worlds Outer Worlds" documentary with Daniel Schmidt for an expanded explanation of "thinking."[100,101]

Credit: Jeff Dahl, GNU Free Documentation License

Given this knowledge, you could suggest the traditional definition of telepathy is not achievable and cannot describe the activity. Perhaps it would be better defined as the process of using our traditional sensory channels (six senses) and as the physical transfer of electromagnetic information.

Conclusions

The purpose of this chapter was to introduce revolutionized scientific knowledge about the transfers of electromagnetic information through the brainwaves synchronicity to the heart's electromagnetic field, discovered by Dr. Rollin McCraty at the Institute of HeartMath (IHM). Our brain has incredible functions, but with the knowledge of both the heart and mind, we free our existence beyond the limitations of the past knowledge. It is the author's position that in the same way that a singer learns to sing a great melody, humanity may discover the emotional reference points that create balanced and cohesive thoughts and a more fascinating human experience.

Furthermore, this chapter is intended not to confirm or deny the traditional narrative of telepathy, but rather to encourage the reader to consider ancient knowledge and new scientific findings for a new perspective. It is the author's position that any definition of a word may be questioned because science is not a religion and great discoveries will always rely upon the revision of outdated definitions as new understandings exceed them.

"Meditation is not so much a technique to master
as it is a re-orientation of the heart; a selfless act
of love and surrender into the mystery and stillness
at the core of our being."

- Daniel Schmidt

Chapter 13

Q&A with Daniel Schmidt
Exploring Inner & Outer Worlds

As our voyage continues, we explore our inner world and review ancient practices that have resurfaced to provide the individual with advanced methods to balance the mind in the pursuit of insight. This chapter is dedicated to the discussion of meditation, yoga and ancient knowledge with award-winning "Inner Worlds, Outer Worlds" documentary filmmaker, Daniel Schmidt.

About Yoga

According to the ancient sages, yoga was a complete universal system, where the mind, body and breath take place in union.

Thus, the word "Yoga" encompassed the whole, not only individual parts, and its entire purpose was spiritual in nature and that remains so today. It is an aid to creating a new perception of what is real, what is needed, and understanding the way of life that embraces both inner and outer realities, ultimately leading to self-realization.[102]

About Meditation

Meditation is an ancient practice that dates back over thousands of years in which an individual trains the mind to reach a higher state of consciousness.

As you silence your mind, you experience the world afar from your thoughts by becoming aware of the here and now - the real address of humankind. You discover that by looking inside yourself, you connect to everything on the outside. Your inner world and outer world meet and there will be no more separation.

This is where insight is found.[103]

About "Inner Worlds, Outer Worlds"

In the ancient traditions, the dharma or "the truth" was always taught freely and never for personal gain or profit in order to preserve the purity of the teachings. As such, The "Inner Worlds, Outer Worlds" documentary film is available for free and all four parts of the film can be found at: innerworldsmovie.com or on the AwakenTheWorldFilm YouTube channel.[104,105]

The film could be described as the external reflection of Daniel's own adventures in meditation. As he came to meditative insights, he realized that these same observations were found over and over in spiritual traditions around the world. He realized that it is this core experience that connects us, not only to the mysterious source of all creation, but to each other as well and that many past scientific geniuses have also come to this conclusion. He and his wife Eva began the process to "Awaken the World" by bringing this ancient knowledge back in our consciousness for the pursuit of restoring balance and harmony on the planet.

Their next film, entitled "Samadhi," is to be released in 2015. It is described as: "The ancient teaching of Samadhi leads to a profound realization of our true nature and a re-orientation of human consciousness. Samadhi cannot be conveyed through words, art or music, yet we vow to convey it."

Q&A with Filmmaker Daniel Schmidt

Q1: You've been on a very interesting journey exploring the world with yoga & meditation. Do inner and outer emotions interfere with your freedom to explore?

A: Emotions, feelings, sensations and thoughts are of course an integral part of a complete human experience. They are only a problem if you identify your "self" with the emotion, thought or sensation. If you can just let an emotion be as it is, without any inner (or outer) contraction or resistance, then it simply becomes more aliveness, more juice for life. It is neither good nor bad in and of itself. The degree to which one can allow everything, including emotions, thoughts, feelings, the breath, inner aliveness, to be as they are, while at the same time being completely present with them, is the degree to which one could be said to be awake.

When the next thought or feeling does not snag your consciousness or trigger unconscious patterns, then you are truly free to explore. You are only free when you are not moved by unconscious motivations. Once you are conscious of an emotion then by definition it is no longer unconsciously moving you, and whatever action ensues is a choice. It all comes down to choice, and a realization that the YOU that is exploring is something unfathomable, and you are actually the thing being explored. You are both the detective and the mystery.

Q2: What is it about meditation that fascinated you? Did you always feel the need to pursue its knowledge?

A: For much of my early life I had a feeling that somehow I was not really awake and that I would eventually wake up. I felt that life was unreal and I wanted to make it more real, but didn't have the tools. This feeling eventually faded and I found myself on a materialistic path that caused me a lot of suffering and brought about deep sickness and imbalance in my mind-body pattern. After years of deterioration that took me to the edge of death, I had an awakening or realization. I understood that the dis-ease was the intelligence of my being nudging me back onto my path. I eventually learned to listen to my inner world and to find balance, seeing the process as a wonderful gift, however painful. It was a process of letting go of everything that was not in alignment with my inner direction, and involved many hours of consciously sitting without resistance to what was unfolding within my being.

My fascination with meditation started not so much as an intellectual interest, but a vague sense of self-preservation, and a desire to find the off button for a pathological mind. Once I realized the mechanism by which I created my own suffering and imbalance, I became excited, and again the feeling returned that I might wake up, and this inner feeling began to guide me. The crazy irony is that as one progresses in meditation one realizes that there is no "I" that awakens. There is no ONE who awakens, literally. There is only awakeness, or consciousness itself, and I am that. Everything else that I refer to as "me" is an obscuration of that truth. As soon as we introduce it into language the truth is lost, and we have a subject/object dualism, which has made the teaching of the truth so difficult.

Q3: It's important to connect with someone, to understand their struggles, but how can you balance empathy whilst protecting oneself from absorbing potentially too much negative energy?

A: My answer to that needs to be prefaced with a certain understanding about duality. The path of meditation leads to Samadhi, which is not an altered state of being, but rather it is our natural state of being. Samadhi is clear consciousness unfiltered by the senses and unmediated by thought. Most humans are in the altered state now, filtering a trickle of energy through the human mechanism, and most identify the SELF with that trickle of energy or phenomena. Samadhi, which is both the seed and the flowering or fruition of meditation, is a union or merging of polarities, of the left and right brain, yin-yang, Shiva and Shakti. It is a merging of the inner and the outer, and of self and other.

If you directly experience yourself AS the other, then there is no labeling of good or bad energy, my energy or another's energy. There is only energy. It is only the mind that wants to label and to protect. Who is really there to protect? I am not separate from the pathological, insane, egoic constructs that are ravaging the world. Everyone necessarily has an ego in order to function as a human, but it is the IDENTIFICATION with the form that is the illusion and ultimately allows the egoic structures to run amok. We feed these ego machines with our consciousness. This is the fundamental paradox; to allow everything to be simply as it is without resistance. When there is no resistance, what is perceived as negative from one point of view, will arise and pass away very quickly.

Q4: Other than meditation, Eva and you have developed different expertise. What is your feeling towards authority and the limits of specialization?

A: For me the most important thing is having a beginner's mind. The moment you think you know, you are truly lost. Eva and I try to dance with the mystery and explore what excites us, and teachings and ideas tend to flow through and around us quickly. Sometimes we use instruments to connect people to their inner world. A gong or a didgeridoo or chanting can be a great way to directly connect someone with their vibratory nature. We don't hang onto rituals or repetitive actions for long, otherwise they might turn into dogma or worn out truths that the mind loves to repeat. As Kierkegaard said, if you name me or label me you negate me. To become an expert or to develop a technique is to hold onto dogma so I prefer to be a beginner each time.

I don't feel a need to be part of any hierarchical group or structure or follow any authority other than my own inner guidance. I am not interested in any part of the Guru game because I know that it only leads to delusion and the building of a spiritual persona. I am fascinated by many of the wisdom traditions and feel that there is much to learn from those great teachers who have come before. But as they say, if you meet the Buddha on your own path you must kill him. You must kill any other Buddha you meet because you will start to follow them rather than the path to your own Buddha nature. But just because you kill him doesn't mean you can't learn a great deal from him. In fact he becomes a true teacher precisely at that moment when he has died for you, if you know what I mean.

Q5: Can anyone's consciousness shape the world alone or is a collective understanding needed to find our own individuality?

A: To awaken oneself IS to awaken the world. Literally. They are not separate. There is no real awakening if it is only this one little human "me." The truth of who you are is vast, connected to all that is. Consciousness not only shapes the world, but consciousness IS the world, or the world happens in consciousness, AS consciousness. I can't be fully awake, fully realized until my fellow humans are awake. If we are all in prison together, I can't truly experience freedom by simply jumping over the wall. I might escape my prison by jumping over the wall and I might experience a kind of relative freedom, from one limited perspective. But I can only taste real freedom, multi-faceted or deeper collective freedom, once the others in the prison are freed as well. Then there can be real celebration, and a new level of heart-connected conscious freedom.

Q6: Richard Feynman, quantum physics professor said, referencing the pursuit of scientific knowledge, "it does no harm to the mystery to know a little more about it. Far more marvelous is the truth than any artists will ever be able to imagine it to be." In his view, we will never know what this world is about but we must have the freedom to explore what fascinates us most.

A: I love Feynman because he was really alive and excited, and he wasn't afraid to look into the shadows of the science world. He could see things that others couldn't because they were still in the collective dream of scientific realism which humanity is still awakening from. Yes, it is the exploration or journey that is important, and we are creative participants on that journey, not just tourists looking at the scenery. The paradox is, as Feynman himself knew, the more you learn about the mystery the deeper it gets.

Q7: What is your view on the relationship between quantum physics and meditation?

A: Einstein described quantum physics as "spooky action at a distance" and I kind of like that description because it is true. It is wildly fascinating, magical, unbound by the usual rules. Meditation takes you to an inner world similarly unbound by logic and thought, but yet, as in quantum physics, there are truths, insights and new landscapes that emerge. Both take you to the threshold at which thought creates a limitation or filter for reality and the point at which the observer and the observed begin to bleed into one another.

Q7: Do you practice transcendental meditation and if so, is it accessible to the average person?

A: I don't like the word transcendental just because it can be confusing and it makes it sound like you are going into some altered state of being or moving out of "normal" life. Meditation is not transcending life, but going deeper into it. Meditation is the simplest and most natural thing. It is only the thinking mind that makes it complex. If I could sum it up, I would say simply observe reality as it is, in the present moment, without resistance, without turning away from your direct experience, and without getting lost in your thoughts.

This self-enquiry is what the Buddha taught, and it is absolutely accessible to all. The most important thing is to inquire into the reality of this moment, and never let the mind become mechanical or latch onto some technique. Experience each breath, each sensation, each moment anew. We get lost in Maya or illusion when we identify our SELVES with the phenomena that is arising, and we suffer when the phenomena changes, if we are attached to it. Many techniques will help you to discipline the mind, to focus on the breath to stay present.

This is great at the beginning because the mind is wandering everywhere and you have to use a bit of effort to make it stay put, but eventually you have to let go of your focus on the technique. It is like playing piano. You learn the notes and the technique at the beginning, but to play really well you have to stop thinking about technique and really feel the music. You have to INHABIT the music, inhabit the playing. In martial arts you learn all kinds of moves, but if you are focused on the technique in a fight you are going to get beaten up. You have to naturally, intuitively respond to each moment, and the technique is there in the background. It is a part of you, but it is nowhere and present at the same time.

Your meditation has to be the same – each moment must be alive, vibrantly alive. Each breath is a fractal, a magnificent snowflake of energy connecting you to the divine source. Your meditation should be alive, dynamic, as you are observing each changing moment. Meditation should not be some grim routine as it is often taught. Your average person will be repelled by a grim routine, and they should be.

Q8: The benefits of meditation are known, but scientifically less understood. How do you think video sharing on the Internet can change this relationship? Can it change the world?

A: Yes, it can change the world. We are at a unique time in history where we have the tools to reach millions of people and to bring the ancient teachings back and to create resonance that will help to open the collective human heart. Science can be a wonderful tool to shed light on some things and to help us make the journey from the head to the heart. At the edges of science we find it pointing back to ourselves.

Bridges must be re-built between science and spirit, or the inner and outer, to show that these worlds are connected, flowing from one to the other. The modern scientist must once again be willing to hold two worlds together, like a Pythagoras or Aesclepius – a techno shaman wizard sage who uses all of his faculties, all of the tools available. Honing our cognitive tool (our mind) through meditation is essential to the investigative process and can help to inform the direction of science. Science will wander aimlessly and will not understand how to benefit humanity until we know and understand how we create suffering on this planet and until the outer world is informed by the intelligence of the inner world.

Q9: Do you believe in some form of Reincarnation and/or multiple dimensions?

A: The whole notion of reincarnation becomes sort of moot when you directly experience that everything is one consciousness. Who reincarnates?...YOU. Who doesn't reincarnate.?...YOU. The real question is "who are YOU?" This is not a question for the mind. The answer must be lived and can only be lived when we shift our focus away from identification with thoughts. Belief is something of the mind. It is unnecessary when there is direct experience of the truth.

The idea of different dimensions can be understood by observing the Metatron's cube figure. If you look at it a certain way you see it as flat, another way you see three-dimensional possibilities. ALL dimensions are like that – simply orienting our consciousness in a certain way to pick out a pattern layer and then staying attuned to that layer.

Q10: Meditation and Pranayama breathing can heal the mind/body/soul but many struggle to find stillness. What advice would you give to those who experience interfering thoughts?

A: Get curious, REALLY curious. The more curious and determined you are to get the root of the mechanism through which the witness interacts with thoughts, the more you will orient your consciousness in correct relationship to thought. Watch your thoughts like a scientist examining a really fascinating experiment. When do the thoughts increase, when do they decrease? Who is thinking these thoughts? If there is someone watching the thoughts, then who is that? Don't push thoughts away, but don't get sucked in by them or identify with them either. Keep consciousness in the middle – the middle way that the Buddha speaks of is to be simply awake, aware without identification or attachment to any arising form.

To meditate is to release control of the breath; to do pranayama is to control the breath. If you continue each practice with single-pointed unbroken intention, then eventually that which moves the natural breath, and the one who controls the breath are realized to be one and the same. When there is Samadhi, there is no difference between Pranayama and meditation because controller and the controlled are one. Be aware of who is breathing and who is being breathed.

Q11: Many believe that a strict vegetarian diet is required to attain a higher self. But considering that animals eat other animals, should humans be held to a different standard?

A: There may be times when a light diet is needed to "lighten up" or to move towards enlightenment. Then there also may be times when one is too light or unrooted, needing grounding, or the body is damaged and a heavier diet is needed. Always listen to the body and the inner world and never make a "one-size-fits-all" assumption when it comes to diet. Your needs may change throughout your life and every person has a unique balance of elements. Maybe as you are able to allow more prana within your system you will need less food. Take direction from the inner world. When people lose their connection to the source, then they create religions or systems to help point the way back to the source. When these religions no longer function, then they create moral codes, rules and laws, but these are poor substitutes for a direct connection to the guiding intelligence of the source.

Q11: Inner Worlds has had great success in reaching its viewers worldwide, what's next for you and Eva?

A: The next film entitled "Samadhi" is in the works. After that the plan is to teach meditation. We simply want to provide the optimal conditions for people to do their own inner investigations. We want to bring the ancient teachings back into the world in an accessible, modern way, using all the tools available. This is what the Awaken the World initiative is about.[106]

Q12: What are you most passionate about?

A: We are most passionate about the awakening of human consciousness on the planet at this time, and the possibility of freeing all beings from the bondage of pathological thinking and identification with material reality.

Namaste!

Conclusions

The purpose of this chapter was to enhance a reader's understanding of yoga and meditation and their potential benefits - from ancient roots to the modern application. These can be tools that enable us to explore beyond what is known and unknown. By way of learning to silence the mind, we become better at listening to the greater sources of wisdom. We believe that educational and healthcare leaders should become more mindful of the importance of yoga and meditation as central element in attaining physical and mental health. It is the author's position that should these practices gain widespread adoption, and be properly intertwined with our modern lives, humanity will achieve a more powerful vibrational frequency and become a more harmonizing force in the Universe.

"Look deep into nature, and then you will understand everything better."

- Albert Einstein

Chapter 14

Q&A with Geneticist Dr. Mae-Wan Ho to Discuss the New Science of Epigenetics

Our exploration continues as we rendezvous and we set sail towards an epic conceptual revolution that begins to recognize the environmental and social effects on gene expression.

As such, this is a chapter dedicated to a Q&A with Dr. Mae-Wan Ho exploring the fascinating new science of epigenetics, the false conceptual revolution of genetic determinism and The Institute of Science in Society (ISIS) magazine. From the Latin word "epi," the term epigenetics refers to (over, outside of, around) the gene.[107]

About Mae-Wan Ho

Mae-Wan Ho, B. Sc. Hon. (First Class) and PhD., Hong Kong University, Director of The Institute of Science in Society (archived by the British Library as part of national documentary heritage), Chief Editor and Art Director of its quarterly trend-setting magazine, Science in Society, is best known for pioneering work on the physics of organisms and sustainable systems, presented in *The Rainbow and the Worm, The Physics of Organisms* (1993, 1998, 2008) and *Living Rainbow H2O* (2012).[108,109]

The importance of structured water for living systems (which she refers to as liquid crystalline water because its molecular organization resembles a solid crystal), figures prominently in her books, both best sellers of the publisher. Her work resulted in winning the 2014 Prigogine Medal.

The Mainstream View of Genetics Determinism Has Been Incorrect

In the mid-1970s, geneticists found exceptions and violations to every principle of classical genetics. They discovered a constant crosstalk between genomes and environment. It was then discovered that feedback from the environment not only determines which genes are activated or turned off, but also where, when, by how much and for how long, as well as marks, moves and changes within the genes themselves.

Despite data accumulated since the 1970s, leading politicians and scientists ignored the evidence and continued to pursue a more convenient theory of genetic determinism that subscribes to the belief that disease is purely a genetic condition. As such, this also served to discouraged many people from questioning social and environmental factors imposed by a ruling class.

The Great DNA Let-Down

When the human genome sequence project was announced in 2000, even President Clinton said it would "revolutionize the diagnosis, prevention and treatment of most, if not all human diseases."

Ten years later, Fortune magazine called the actual findings, "The great DNA let-down."[110]

David B. Goldstein at Duke University, a leading young population geneticist known partly for his research into the genetic origins of the Jews, said: "It's an astounding thing, that we have cracked open the human genome and can look at the entire complement of common genetic variants, and what do we find? Almost nothing. That is absolutely beyond belief."[111]

The New Science of Epigenetics

It turns out that environmental and social conditions play a critical role in gene expression. As such, a new discipline of epigenetics is emerging to study inheritance "outside" genetics, i.e., not due to the DNA sequence of the genome. Epigenetics reveals how proteins are assembled; how specific genes can be expressed or not; and how environmental and social pressures can even recode the human genome.

Genes Alone do not Determine Gene Expression

According to Dr. Mae-Wan Ho, genome-wide scans for linear genes responsible for common diseases or intelligence have completely failed, while environmental epigenetic changes perpetrated on the genome are remarkably predictable. A single gene for "x" – has been decisively proven incorrect. Only in extremely rare circumstances does it hold true.

As such, genetic determinism, which has been promoted by the mainstream for hundreds of years, has turned out to be the most significant false conceptual revolution of modern times.

Individual Experiences

There is definite causation between genes and environment, which means that new genetics and environmental effects are inseparable. Evidence of the inextricable entanglement between the organism and its experience of the environment is forcing us to rethink not only genetics, but also evolution.

Q&A with Geneticist Dr. Mae-Wan Ho

Q1: Is scientific authority respected for the wrong reasons? Have we marginalized key figures of authority to fancy titles instead of the quality of solutions provided? As such, has authority abused this power to obstruct broad and honest knowledge sharing?

A: The scientific establishment for too long has cloaked itself in obscure jargon, high-sounding language and other trappings to appropriate authority it does not deserve, at the same time misinterpreting data to undermine the public's understanding of critical issues. It becomes all too easy for the powers that be to exploit this false authority for political and economic ends, for they are natural allies in the matter of projecting pretensions. ISIS is dedicated to exposing those pretensions in the interest of bringing real freedom of choice, equality, and democracy to people. We believe in promoting critical understandings and knowledge for all people.

Q2: In light of the phenomena called epigenetics, did the mainstream media and many leading scientific voices create a highly profitable but false conceptual revolution around genetics determinism?

A: Yes. Those in favor of control and profit invariably favor things that can be owned and sold for profit. They also favor the status quo, especially the idea that life is just a genetic lottery and those who succeed are the ones with the best genes, and therefore have an almost god-given right to rule.

The new genetics and indeed the new science of the organism elaborated in my books emphasize process, circular causation and require active participation in shaping one's own development and evolution.

It completely challenges the old paradigm that has dominated science and society for close to a hundred years. I have spelled out the implications in a recent article, as follows: "As regards implications for social policies, we already have a great deal of knowledge on how social deprivation, psychological stress, and environmental toxins can have dire effects on us and our still unborn children and grandchildren while social enrichment, caring environments and cognitive and physical exercises, and stress reducing mind-body techniques can have beneficial effects on infants, children and adults alike. The implications on the appropriate interventions for health, education and social well-being are clear."

See below for an expanded view:

- Epigenetic Inheritance – What Genes Remember[112]
- Epigenetic Toxicology[113]
- Caring Mothers Strike Fatal Blow against Genetic Determinism[114]
- Living, Green and Circular[115]
- Evolution by Natural Genetic Engineering[116]

Q3: Why organisms are, among other things, so sensitive to electromagnetic fields and microwaves of very low intensity, very weak fields, such as those from high-tension power lines and mobile phones?

A: The short answer is that living systems are quantum coherent and depend on very weak and precise electromagnetic signals for intercommunication, not only between molecules, cells and tissues to coordinate biological function, but also between the organism and the electromagnetic rhythms of the earth and stars to which organisms are attuned through billions of years of evolution. External electromagnetic signals of the strength, diversity and saturation we have today are late arrivals to the living environment. They invariably interfere with the electromagnetic signals that support and organize life.

Q4: You have written several articles about structured water on the ISIS website and have thoroughly discussed it in your most recent book "*Living Rainbow H2O* (2012)." How important is liquid crystalline (structured) water to the new science of organisms?

A: The importance of water in living organisms associated with membranes and other interfaces has been recognized by many pioneers of biochemistry, especially Albert Szent-Györgyi more than 50 years ago.

The contribution made in my laboratory was in discovering that organisms and cells are liquid crystalline and coherent to a high degree, even quantum coherent, thanks to the water at the interface.

This water is itself liquid crystalline and essential to the functioning of biological molecules. Without water, macromolecules cannot work at all, water makes all the difference; it enables macromolecules such as enzymes to function as quantum molecular machines at close to 100 % efficiency. If not for that, we would literally burn out with all the heat generated before you could say Christopher Robin!

To give a simple illustration, computers have reached the limit in miniaturization simply because the electronic reactions are so densely packed that far too much heat is generated. The density of reactions in our cells is at least a billion fold that of the best computers, and we operate with a tiny fraction of the energy required by the computers. That alone is sufficient to convince us that organisms need something close to quantum coherence.

Liquid crystalline water is the key in another way; it is water in the excited state, as Szent-Györgyi pointed out, ready to be split into protons and electrons with energy from sunlight. If not for that, photosynthesis would be impossible. Photosynthesis by green plants, algae and blue green bacteria provides energy for fuelling life itself to support oxygen-breathing organisms like us, and electricity for animating life.

In addition, water in tight nano-spaces, such as those existing in cells and tissues, are found to adopt entirely new quantum states that are superconducting and superfluid at room temperature and pressure, which could be crucial for intercommunication and for maintaining water and electrolyte balance. This is all very exciting in terms of potential applications in photonics and electronics, not to mention holistic health, water purification, new quantum devices etc.

Gerald H. Pollack's important contribution is in discovering (or rediscovering) that interfacial water can exist in the macroscopic domain. His lab shows, for the first time, that interfacial water excludes solutes, reacts to light and result in charge separation. In addition, it gives directed flow and other interesting phenomena.

We don't quite understand interfacial water as yet, but big advances are being made, and bigger advances are to be made still if the conventional communities weren't so irrationally afraid to investigate these phenomena. Water, it seems, can be structured in the large, as well as in tight places, inside cells and tissues. Water is indeed the basis of a new science. Basic biology and health are poised for a revolution.

See below for an expanded view:

- Water Structured in the Golden Ratio[117]
- The Importance of Cell Water[118]
- DNA Sequence Reconstituted from Water Memory?[119]
- New Age of Water[120]

Q5: Why is there a lack of global consensus within the scientific community on genetically modified food (GMOs)?

A: There are several reasons. Many scientists working to create GMOs have a conflict of interest that makes it all too easy not to consider the risks, and to be narrowly focused on the presumed benefits. It is also partly down to the specialized training they received. And most importantly, there is little or no funding for research on risks of GMOs. Some scientists and non-scientists are funded by the biotech industry to be mouthpieces; most of them, even the scientists don't even know enough about molecular genetics, and may genuinely believe what the 'experts' tell them.

Biotech corporations also make it very awkward indeed for risk research to be carried out, they withhold material as well as raw data (with connivance from the regulators), and go to great lengths to discredit and suppress evidence of harm from GM feeding studies, as witnessed in the recent machinations to unilaterally 'retract' a study from a scientific journal a year after it has gone through a proper review process. See "Retracting Seralini Study Violates Science & Ethics."[121]

Still, we do have nearly 300 scientists declaring no consensus on GMO safety. The scientific debate about GM foods is by no means over. In fact, a real consensus is building up among scientists, UN agencies and consumers worldwide based on real evidence that small agroecological farms are the way ahead in tackling climate change and providing food security. They are currently the major sector in the global food market, producing more than half of the food consumed in the world.

See below for an expanded view:

- Scientists Declare No Consensus on GMO Safety[122]
- Global Status of GMO and Non-GMO Crops[123]
- Why GMOs Can Never be Safe[124]
- Evolution by Natural Genetic Engineering[125]
- Horizontal Transfer of GM DNA Widespread[126]

Q6: To the Hindu-Yogi, meditation has long been professed to have healing abilities. Has the new science of epigenetics begun to support this belief? Can our minds change gene expression with the practice of meditation?

A: There is no doubt that mind-body exercises of all kinds are capable of changing gene expression. It breaches the mind body divide that's the hallmark of mechanistic Western science. Like people from other great cultural traditions in the world, I have never subscribed to the mind-body divide. I would wager that mind-body exercises could mark and change genes as well, and evidence for that would emerge before long.

- How Mind Changes Genes through Meditation[127]
- No Genes for Intelligence[128]

Q7: You've mentioned that there is abundant evidence which points to the enormous potential for improving intellectual abilities (and health) through simple environmental and social interventions. Could you provide us with a few examples that fascinate you most?

A: I have written a review on (How to Increase the Brain Power and Health of a Nation)[129] the potential for improving intellectual abilities ranges from providing adequate early nutrition, to education and enrichment programs (including martial arts and mindfulness exercises), memory training, and physical exercises, the latter two are effective, not only for the young but especially also for the elderly.

The importance of good nutrition for young children is so clear that interventions aimed at eliminating food insecurity and micronutrient deficiencies – easily within the means of all developed nations – should be given top priority in both developed and developing nations. Studies in Central America and Panama showed that supplementary feeding of infants and young children with drinks that provide energy and micronutrients, or with added protein, both resulted in significant increase in cognitive development and school performance through to adolescence.

A meta-analysis of a large number of studies on older adults showed that aerobic exercise for the elderly is important for maintaining IQ, and reducing the likelihood of Alzheimer's disease, especially for those over 70-years-old. There is also evidence that exercise increases brainpower through generating more neurons.

Not so long ago, neurobiologists and the general public believed that we were born with the neurons we would have for the rest of our lives. That dogma was overturned in the 1990s. New neurons are continually generated in two regions of the brain.

Q8: You have been a strong critic of the Neo-Darwinism theory of evolution by natural selection and instead have favored Lamarck's theory of transformation arising from the organism's own experience of the environment. In your opinion, what could be wrong with Darwin's theory of evolution?

A: The neo-Darwinian dogma that evolution occurs by the natural selection of random mutations is just plain wrong. It is contrary to all the evidence in molecular genetics accumulating since about the mid-1970s and especially since the human genome has been sequenced. Not only is mutation non-random, the experience of the individual marks and changes genes influencing their own development as well as the development of their children and grandchildren. This does support so-called Lamarckian theory of the inheritance of acquired characteristics. Neo-Darwinism is an ideology pure and simple. It glorifies competition and competitiveness and instills a fatalistic paralysis in people. The new genetics on the contrary tells us that we must take an active role in shaping our own destiny as well as looking after the wellbeing of future generations.

More generally, we must give up the obsolete mechanistic science for the science of the organism, which accords a central role to symbiosis and cooperation in place of competition and exploitation in the survival and evolution of the fittest. See *The Rainbow and the Worm, The Physics of Organisms* and *Living Rainbow H2O* for more details.[130,131]

Conclusions

The purpose of this chapter was to establish genetic determinism as one of the greatest false conceptual revolutions of our time while introducing the new science of epigenetics. This new science provides fascinating information as to how environment and socialization impact gene expression and affect human health. It is the author's position that we must seek greater balance and cohesion between past, current and future generations as it relates to the impact of social and environmental pressures on each other. The healthcare industry will always have a significant position in society but it requires a significant cultural transformation with broader considerations and supporting research on the social and environmental factors on gene expression. We believe that the study of epigenetics will become a central part of our consciousness and enable a more dynamic world of possibilities to emerge.

"Our lives begin to end the day we become silent
about things that matter."

- Martin Luther King, Jr.

Chapter 15

Q&A with Dr. Gerald H. Pollack, Conceptual Revolutions and the Institute for Venture Science (IVS)

We have now arrived at the base of a great mountain. Our objective is to bring everybody to the mountaintop. To accomplish this task, we must pursue selfless leaders with an honest passion that ensures each next step taken is anchored to the scientific truth as currently understood and without regard for old and outdated theories.

This chapter and interview with Dr. Gerald H. Pollack is dedicated to the discussion of his groundbreaking re-discovery of structured water, conceptual revolutions and the Institute for Venture Science (IVS). First let's take a look at critical issues that impact scientific funding today. Check out the website at www.theinstituteforventurescience.org.[132]

Dr. Gerald H. Pollack is a professor of bioengineering from the University of Washington faculty and is the Editor-in-Chief of the journal WATER. He is best known for pioneering work on the new science of water presented in *Cells, Gels and the Engines of Life* and in his most recent book, *The Fourth Phase of Water: Beyond Solid, Liquid, and Vapor.* Both are available for purchase.[133] For a complete list of his publications, please visit the Pollack Laboratory website.[134]

Of note, Dr. Pollack received the highest honor the University of Washington in Seattle can confer on its faculty for the discovery of structured water.[135]

Conceptual Revolutions

Before we began the interview, Dr. Pollack noted, that over recent decades, scientific breakthroughs that support conceptual revolutions have been few and far between. He's not talking about technological revolutions such as the Internet and the cell phone, which are supported by the private sector for profit. He's talking about conceptual revolutions that significantly change our outlook on the world.

Take, for example, the problem of cancer. More than 40 years ago, President Nixon declared war on cancer. Yet, today's therapies show only incremental improvement over that period of time. It seems, winning the war on cancer cannot be legislated simply by declaring that it is important. Nurturing revolutions demand the scientific freedom to challenge authority, the ability to revisit old paradigms and the encouragement to follow the evidence wherever it may lead.

Why does our modern system work for incremental science but struggle with breakthrough scientific investigation? According to Dr. Pollack, "This is because current grant agencies are not set up to deal with the most critical obstacle to realization: the reluctance of the scientific community to entertain ideas that challenge their long-held views."

The Modern Science Funding Mechanisms Favor Publications

Since the time when federal granting agencies were created, more than half a century ago, the outpouring of scientific data has been staggering. The number of papers published over the past three decades exceeds the number of papers ever published in history. For example, in the United States the National Institutes of Health and the National Science Foundation receive ~US $40 billion dollars and Defense budgets, receive US $60 billion dollars annually for various research projects. As such, you have almost US $100 billion dollars invested in scientific research. Nevertheless, there are far fewer conceptually revolutionary scientific discoveries as compared to a hundred years ago when Einstein and others were around. Why? Dr. Pollack further explains:

"Scientists of a hundred years ago were driven by passion. Their compelling desire to understand nature propelled them toward discovery. Today's scientists differ. Science no longer comprises a passion for few, but a profession for many, much like law, medicine, etc. Some scientists retain passion to understand, but many are driven principally by the desire to succeed in their professions. That's a different kind of goal. We now measure success less in terms of discovery but more in terms of the numbers of publications, the size of laboratories, the amount of grant money received, etc. Those "achievements" are the modern carrots dangling from the ends of sticks. Hungry scientists competitively seek those carrots of recognition. Those appetites differ from the hunger for truth."

Well-Funded Scientists are Given Power to Suppress Opposing Views

During the review of grant proposals, science administrators seek the most established scientific leaders – the leading proponents of the status quo. Any applicant challenging the views of those leaders rarely succeed and therefore, existing paradigms persist even if inadequate. Risk aversion has become the norm. Therefore breakthroughs can rarely and barely emerge. According to Dr. Pollack, "most scientists with revolutionary ideas keep it secret because of the risk of developing a reputation as being the kind of person to stay away from."

Furthermore, it is no longer a matter of convincing a few, but of convincing huge masses whose collective specialized influence in maintaining the status quo have control over the funds available.

As a result, only a few meaningful scientific breakthrough ideas ever come to realization.

The Bold New Proposal: The Institute for Venture Science

Even the funding agencies themselves recognize the breakthrough problem and, that something needs to be done to truly pursue transformational science. However, pursuing such a bold investment strategy for scientific progress requires a vehicle designed specifically to meet the terrain.

This vehicle needs to differ from that of traditional science-funding organizations and push towards more relevant models that offer greater flexibility and the ability to revise old paradigms in the pursuit of scientific truth. As such, The Pollack Laboratory and others are proposing a ~$US10 billion dollar endowment to fully fund an independent scientific institution, unencumbered by past philosophies and procedures called the Institute for Venture Science (IVS).

After a ramp-up period, the Institute is expected to operate on a $1B per year budget, drawn from a permanent endowment. Dr. Pollack is confident that within the first 10 years, the Institute for Venture Science can make meaningful and revolutionary discoveries in all sciences.

With success of the Institute, that endowment could be increased.

Funding Promising Ideas that Challenge Conventional Thinking

The Institute for Venture Science (IVS) is conceived to fund high-risk non-traditional scientific inquiries that are likely to yield important breakthroughs. To achieve the needed action, they'll invest in promising ideas that challenge tired, worn-out paradigms.

The number of unconventional schools of thought around today is surprisingly large. According to Dr. Pollack "Those who take the trouble to look will find meaningful alternatives distributed throughout practically all domains of science, many of them rich with promise. All but a few are ignored or repressed by the prevailing orthodoxy. These unconventional schools of thought represent potentially ripe fruit, waiting to be plucked."

The IVS will receive proposals worldwide from all realms of science. Scientists from outside the respective proposals' area will judge their merit, thereby minimizing biases or self-interest. A pool of the most highly rated plans would then receive funding.

The IVS will Fund the Idea, Not Just the Person

The IVS will invest in groups of scientists who independently pursue the same unconventional approach to an entrenched way of thinking or an intractable problem. The Challengers and the status-quo position will then compete on equal footing and, the better of the two approaches will prevail.

Culturally, the institute will nurture the actions of challenge and review of old paradigms to enable a more open, coherent and broad scientific debate.

According to the IVS website "Evaluation of these challenge proposals will involve debates in which proponents of alternative schools of thought argue their cases against the respective orthodoxies. A panel of scientists will judge the outcome. Panel members would be close enough to the field to be able to evaluate the material, yet removed enough from the field in order to avoid conflict of interest. New paradigms judged to be the most plausible and most potentially far-reaching will be selected competitively for funding."

"Funding should be enough for, say, a dozen groups to carry forth on the same theme. Duration will need to be sufficient for building traction; yet it should be short enough to avert wasteful spending. With such a funding program in place, promising new paradigms should be quickly elevated to competitive status (a process begun with the initial web debate). Argued in a civilized manner, each paradigm's strengths and weaknesses will soon become obvious, and the superior one should quickly emerge as a realized revolution. Instead of hanging on the vine to wither, the low-hanging fruit will have been expediently harvested."

The Funding Campaign

The IVS is now seeking private sources for funding from the many billionaires who have signed the Giving Pledge by Bill Gates. Many of them are likely to be soon donating substantial amounts to various charities. As of 2011, 69 billionaires had joined the campaign to give 50% or more of their wealth to charity.

The Institute will provide an opportunity for donors to leave a lasting mark. The scientific revolutions anticipated from this new structure will enrich the world in ways that cannot yet be conceived.

Check out the website for The Institute for Venture Science and help by sharing this information to create global awareness about this crucial project.[136]

"Education is the kindling of a flame,
not the filling of a vessel."

- Socrates

Q&A with Bioengineering Professor Dr. Gerald H. Pollack

Q1. Is authority respected for the wrong reasons? Have we marginalized scientific authority to a fancy title and language ... instead of the quality of solutions provided? As such, does authority take advantage of this position to pursue self-interests and avoids solutions that challenge status-quo understandings?

A: Yes. Today's culture emphasizes self-interest, and science has come to subscribe to those cultural norms. Recall the expression, "Power corrupts and absolute power corrupts absolutely." Authoritarian power increasingly dominates the scientific culture.

Those who challenge prevailing views commonly suffer ostracism. Some of them acquire labels as "denialists," implying that their positions are as outrageously extreme as those who deny the Holocaust.

Such abuses of power are hardly new. What's new is that those in power dominate the grant-review bodies. They have the power to choke the funds from challengers, which effectively suppresses challenges. Without allowable challenges to prevailing views, science runs the risk of stagnation.

Q2. From our viewpoint past imaginative geniuses embraced all bodies of knowledge in their development and capacity to solve problems. What is your feeling towards specialization? Does a deficiency in broad knowledge and interests prevent scientist from facing new realities?

A: Defining the boundary between physics, chemistry and biology is practically impossible. What is biology without chemistry? Does atomic structure lie within the discipline of chemistry, or physics? And, don't atoms play roles in biology? By contrast, today's scientists specialize: Actually, they super-specialize. Scientists become world experts in the narrowest of fields. By so doing, they lose sight of potential contributions from other spheres of understanding. Today's scientists dwell on the tiny peripheral branchlets of the tree of knowledge. They haven't the time to consider whether the limbs supporting those branches are sound. Sometimes fresh new growth requires major pruning.

Q3. What is your thought on a better educational system? Can the Institute for Venture Science play a role in keeping educational programs up to speed?

A: Our long-term plan emphasizes educational breadth. One famous scientist remarked that the ONLY courses that taught him how to think were his courses in art and philosophy. Science courses merely dumped information, the students treated as empty vessels waiting to be filled. A better plan educates students to think. It provides broad underpinnings and confers the confidence to build on those underpinnings and the willingness to challenge even long-standing views that don't make sense.

Q4. From our view, we seem to live in world where the majority of the world's citizen's wake up determined to maintain their social economic standing by making others less competitive with dishonest and/or misleading information. Can this institute help to create an environment of more honest observations by better scrutinizing both the mainstream and challenging views?

A: Absolutely. By funding promising ideas that challenge mainstream thinking, the IVS creates an environment that emphasizes truth seeking. If we succeed, then that truth-seeking ethic could begin permeating the rest of the culture. Those who may consider donating funds to the Institute will be endowing a cultural movement oriented toward honest truth seeking and away from self-aggrandizing. They'll be funding a cultural shift.

Q5. In light of the phenomena called epigenetics – and the knowledge that a gene for "X" is false and dangerously misleading – did mainstream media and science create a highly profitable but false conceptual revolution around genetics? If so, how can the IVS minimize such false conceptual revolutions?

A: Yes, mainstream media surely bears some responsibility for perpetuating the illusion that genetics will solve all problems. That revolution seems to be sputtering out of gas. Even some in the genetics community are coming to realize that genetics research receives a disproportionate fraction of scientific resources. Whether the IVS can turn the tide on that and other such illusory revolutions depends on how successful we can become. If we can precipitate genuine revolutions – our main goal – then the public may begin shifting focus from what the media tells them, to what's real. Who knows? The excitement of genuine revolutions may one day permeate the media.

Q6. Water is very prevalent in all life forms. To the Hindu-Yogi, water is Nature's colossal Remedy that contains prana – the "Vital Force." With that, would you suggest that structured water (EZ) contains more of this Vital force?

A: Yes. That's quite possible. EZ (fourth phase) water contains potential energy. Ample evidence shows that this energy gets put to use in our bodies. One colleague from Saudi Arabia was so taken by this possibility that he suggested EZ water as the basis of the "human soul," citing the Koran as basis. When he went so far as to suggest that we write a joint paper on that revelation, I told him that I had enough trouble convincing dyed-in-the-wool colleagues about water's fourth phase (it was early on), that espousing any connection with prana, vital energy, religion, or the human soul might prove fatal to our attempts. On the other hand, more and more I come to realize the wisdom of the ancients. I've seen evidence that waters taken from the Ganges, from Lourdes, and from similar sites bear clear physico-chemical differences from run-of-the-mill water. We've begun experimental tests to determine whether some of those waters can actually reverse pathologies, as touted.

Q7. Regarding rising water; the conventional views as to the significant mechanism behind the ascent drawing force of water against gravity through the trunk of trees is still widely controversial. What does this imply with regards to the organized structure and electrical potential (negative charged) of EZ water?

A: Electrical potentials pervade nature. My recent book, *The Fourth Phase of Water: Beyond Solid, Liquid and Vapor*, explains how and why.[137] Of particular note with regard to water's ascent in trees is our laboratory's finding of "spontaneous" flow through hollow tubes. We found that the flow is powered ultimately by energy received from the environment, e.g., electromagnetic energy including sunlight. Extra incident light produces faster flow. That same energy-driven flow is a good candidate for driving the flow in plants and tall trees and something definitely worth testing.

Q8. Many people practice natural ways to purify water, even offering prayer/intention directed to possibly alter its structure. Are these ways beneficial?

A: Abundant evidence shows that water absorbs energy from the environment. If that energy contains information, then water can theoretically receive information, perhaps store it, and possibly even re-radiate that information. Published, peer-reviewed evidence supports all of those notions, although most scientists remain unaware of the evidence. This is a frontier issue, crying out for serious study.

Q9. The discovery of structured water is likely to hold many applications and touch many aspects of our lives, but which application most fascinates you?

A: The applications do seem countless. This creates a problem akin to eating dessert: having one hot-fudge sundae after a meal can be wonderful, but 15 can be an overdose. My attention shifts continually from one application to another. Right now, two dominate. The prospect of information storage, as mentioned above, could have immense implications for nature, especially for biomedicine. The EZ's semi-crystalline nature provides a potential substrate for storage, much like other information-storing crystals.

The second application that fascinates me is health: EZ water fills our cells. Deficiencies of EZ water lead to dysfunction, i.e., proteins cannot fold properly when sub-normal amounts of EZ water envelop them. The mere act of rebuilding EZ water, by simple, even traditional mechanisms, can possibly heal. That's an exciting prospect that we're gearing up to test.

Q10. You have said that meaningful, paradigm-challenging ideas in diverse fields exist in abundance. Other than your current work, which discovery intrigues you most?

A: The reason I'm reluctant to answer that question is that a mere mention of a topic with potential paradigm shifting character will inevitably invoke reflexive negative responses. No proposal can be taken seriously in the absence of rationale, and this venue offers no opportunity to elaborate. If I were to suggest, for example, that electrical charge powers bird flight, you'd probably conclude incipient lunacy. If I mentioned a radically new model of the atom that better explains nature, you might be tempted to ring up the psychiatric rescue squad. I've seen dozens of meaningful proposals in diverse fields. Unless the underlying rationale can be properly presented, these proposals will make no sense at all. For this reason, I'm unwilling to detail any specifics. On the other hand, the IVS website will contain a representative list of potential paradigm-shifting ideas collected over the full scientific spectrum.[138]

Conclusions

The purpose of this chapter was to discuss the critical issues that impact global scientific funding today in the quest for scientific breakthroughs that can change your outlook on the world. While we have had many technological revolutions, we have seen few conceptual revolutions in modern times. As Dr. Pollack explained, this outcome is the result of the current grant system which, "is not set up to deal with the most critical obstacle to breakthrough science; a reluctance to entertain opposing views that challenge long held belief." Even the battle to cure cancer cannot be secured without a total willingness from the scientific community to follow the evidence wherever it may lead. It is the author's position that the Institute for Venture Science as proposed by Dr. Pollack has a worthy and novel structure to address the terrain to overcome these obstacles. Everything in science can be questioned because it is not a religion and the IVS can be that important organization that supports widespread promising ideas to challenge traditional wisdom.

"Power doesn't corrupt people,
people corrupt power."

- William Gaddis

Chapter 16

Global Warming & Climate Engineering

We end our amazing journey as we skydive back to our home base. Understand, there are many things that you cannot control when you jump out of a plane, but everyone is entitled to a primary and backup parachute. Although, you were given these items by the powers that be, you must always repack your bag to ensure all precautions have been taken.

This chapter is dedicated to the discussion of the global warming issue and to the dangers of property rights qualifications in environmental and social decision making processes.

We know the current cycle of global warming is mainly a problem of too much carbon dioxide (CO2) in the atmosphere – which acts as a cover, trapping heat and warming the planet. The fossil fuels we burn for energy – coal, natural gas, and oil – and the loss of forests due to deforestation, greatly contributes to the excess carbon in the atmosphere. At the same time, certain waste management and agricultural practices aggravate the problem by releasing other potent global warming gases, such as methane and nitrous oxide.

CO2 survives in the atmosphere for a long time, up to many centuries, so its effects are compounded over time. To put this in perspective, the carbon we put in the atmosphere today is essentially free riding on the tax dollars of future generations.

According to work published in 2007, the concentrations of CO2 and methane have increased by 36% and 148% respectively since 1750.[139] Furthermore, these intensities are much higher than at any moment during the last 800,000 years, the period for which reliable data has been studied from ice cores.[140,141]

The scientific evidence is clear enough, there is virtually no debate: an overwhelming majority of climate scientists agree that global warming is taking place and that climate change is largely caused by human activities, but still some industry players are unwilling to face this reality.

Instead, a vocal minority of special interests has proposed to delay action on climate change by releasing additional pollution into our atmosphere. How?

What we are about to discuss should be seen as a clear example of the dangers related to property qualifications rights in the legislation and decision making processes of political leaders.

Reducing Emissions

According to a 2005 study, the top 5 annual greenhouse gas emissions were as follows: electricity and heat 24.9%, transportation 14.3%, industries 14.7%, agriculture 13.9%, and land use changes 12.2%. Between 1972-2012, on a percentage share of cumulative energy CO2 emissions, the United States leads all nations at 26% of total emissions while European Union members and China are not far behind.[142]

In 2006, China over took the United States as the world's largest emitter annually. Now – over 82% of climate scientists believe that actions must be taken to combat climate change, to cool the planet, by reducing emissions.[143]

Much of the discussions amongst scientists have been centered on replacing fossil fuels with renewable energy, halting deforestation and adopting sustainable farming practices. In the end – the solution is likely to be a mixture of these different initiatives.

While nations argue on what to do - one specialized Harvard professor recently shocked the world by proposing climate engineering as a temporary solution to our problems. This is what happens when private interests limit possible solutions and experimental questions, in addressing our environmental problems. This is the parachute that many want to give you!

Climate Engineering: Stratospheric Sulfate Aerosols

In 2013, David Keith released a book, *A Case for Climate Engineering*, which details a highly controversial strategy for slowing climate change.[144] In this, Keith suggested that by spraying sulfate or sulfuric acid aerosols into the atmosphere, a reflective shield could be created to block sunlight and buy more time for humanity to curb emissions, i.e. increasing atmospheric pollution. Delivery methods would be by artillery, aircraft or balloons.[145]

To be effective, these aerosol programs would contain very small particles of aluminum, barium and strontium to temporarily reflect sunlight and these hard metals would then return to surface after rainfall.

Overall, this strategy also known as Solar Radiation Management (SRM) creates a global dimming effect, which is not only a climate engineering consequence. It occurs in normal conditions, due to pollution from natural phenomena such as emissions from volcano eruptions and major forest fires.

Known environmental consequences of these programs include ozone depletion, severe droughts and agricultural damage.[146]

Whether you believe in global warming or not – it's critical to recognize that a strategy to block sunlight and cool down the planet is actively being pursued. And, it's likely to have many negative health effects on an unknowing population.

Only a "Grace Period" and Limited Effects

Scientists who propose introducing these additional sulfate particles or aerosols into the stratosphere acknowledge that this strategy would only provide a "grace period" of up to 20 years before major cutbacks in greenhouse gas emissions would be required.

Furthermore, these methods do not reduce the greenhouse gas effect and thus, do not address problems such as the ocean acidification caused by CO2.

Broad Considerations Needed Before Reaching Justifiable Conclusions

David Keith, Harvard University professor and president of Carbon Engineering, based in Calgary, Canada, confirmed in a 2012 updated documentary "Look Up," that no detailed tests have been conducted to assess the full spectrum of global risks to human health associated with releasing these additional toxins into the atmosphere.[147]

The reality is that Mr. Keith and many of his super specialist followers simply do not understand the entire system well enough to promote these broad experiments onto a mostly unknowing population.[148,149,150,151]

Epigenetics

As previously discussed, in the 1970s it was discovered that feedback from the environment not only determines which genes are activated or turned off, where, when, by how much and for how long, but also how it marks, moves and changes the genes themselves. In addition, an individual's exposure to environmental toxins can be passed on to future generations.

There is definite causation between genes and environment (an individual's experience). This means that new genetics and environmental effects are inseparable.

Evidence of the inextricable entanglement between the organism and its experience of the environment should force legislators to rethink the toxins we release into the air. For a review of epigenetics, return to Chapter 14.

"Look Up" Movement

While we are certain that some scientists have conducted small-scale experiments, several whistle blowers (such as Kristen Meghan, Ex-Military)[152] and investigative journalists are now saying that large-scale programs are ongoing. Many within this movement believe that since GeoEngineering is completely unregulated, private interests have already deployed large-scale programs in an attempt to block sunlight and delay the effects of global warming.

We were astonished at how much anecdotal evidence (videos and images) you can find on the Internet on stratospheric sulfate aerosols. So "Look Up." You may be surprised at what you find.

Conclusions

The purpose of this chapter is neither to confirm nor deny the ongoing speculation about stratospheric sulfate aerosols program activities. We provide this chapter as final and timeless example of the dangers related to property qualifications rights within the legislative and decision-making processes of political leaders.

As a consequence of the narrow interests of many of those in power, we find it particularly fascinating that so much burden of thought has been invested in various solar radiation management techniques despite the fact that it does not address the root cause of our environmental problems and considers so few of the potential negative biological and ecological consequences on these generations and the next.

As Plato explained, "Ethics" is not inherent in rhetoric and that Rhetoricians are those who "would teach anyone to learn oratory but without expertise in what's just." It is the author's position that we should always remember that we cannot cheat ourselves to building a more advanced and insightful human experience; we must do the work.

For our technologies to prosper – higher quality conversations about the full effects of our actions must be pursued for nature cannot be fooled. Everything is connected.

About the Author

Adam B. Dorfman has a mix of professional and entrepreneurial experience as well as an MBA from the University of Toronto in Investment Banking and a Bachelor of Commerce from the University of Ottawa in International Management. Most recently, he founded the website ConceptualRevolutions.com to honor the work of scientists who have made scientific discoveries and film makers who have pursued the philosophy of broad knowledge and in doing so have offered significant insights to a more connected human existence. He is also fluent in English, French and has conversational proficiency in Spanish.

About Relentlessly Creative Books

Relentlessly Creative Books™ offers an exciting new publishing option for authors. Our "middle path publishing™" approach includes many of the advantages of both traditional publishing and self-publishing without the drawbacks. For more information and a complete online catalog of our books, please visit us at RelentlesslyCreativeBooks.com or write us at books@relentlesslycreative.com.

For readers, join our online Readers Group and enjoy free eBooks, sneak previews on new releases, book sales, author interviews, book reviews, reader surveys and online events with Authors. Register at RelentlesslyCreativeBooks.com.

Endnotes

[1] https://www.youtube.com/watch?v=qmL4CeTENtw&list=UUmzsYSXn3zdCsLS063oCt8w
[2] https://www.youtube.com/watch?v=qmL4CeTENtw
[3] http://www.innerworldsmovie.com
[4] https://www.youtube.com/watch?v=qmL4CeTENtw&list=UUmzsYSXn3zdCsLS063oCt8w
[5] http://en.wikiquote.org/wiki/Richard_Feynman
[6] http://en.wikipedia.org/wiki/Visible_spectrum
[7] http://movies.nationalgeographic.com/movies/mysteries-of-the-unseen-world/about-the-film/
[8] http://www.wikiwand.com/en/Hearing_range
[9] http://www.elephantvoices.org/elephant-communication/acoustic-communication.html
[10] http://www.dolphincommunicationproject.org/index.php?option=com_content&task=view&id=1119&Itemid=285
[11] http://newswire.rockefeller.edu/2014/03/20/sniff-study-suggests-humans-can-distinguish-more-than-1-trillion-scents/?output=pdf
[12] http://www.thenakedscientists.com/HTML/content/interviews/interview/1000650/
[13] http://en.wikipedia.org/wiki/Sense
[14] https://www.youtube.com/watch?v=oS6FGzh3ygw
[15] https://www.wikiwand.com/en/Extremely_low_frequency
[16] http://www.wikiwand.com/en/Schumann_resonances
[17] http://www.wikiwand.com/en/Alpha_wave
[18] http://www.wikiwand.com/en/Rütger_Wever
[19] http://www.i-sis.org.uk/FOI3.php
[20] http://en.wikipedia.org/wiki/Colony_collapse_disorder#Electromagnetic_radiation
[21] http://www.nature.com/scitable/blog/green-science/global_crisis_honeybee_population_on
[22] http://www.wikiwand.com/en/Magnetoception
[23] http://www.wikiwand.com/en/Cryptochrome
[24] http://www.nytimes.com/2011/06/28/science/28magnet.html?_r=0
[25] http://www.bbc.com/news/science-environment-13809144
[26] http://cymascope.com/cyma_research/oceanography.html
[27] http://www.nonoise.org/library/epahlth/epahlth.htm
[28] http://www.wikiwand.com/en/Stradivarius
[29] https://www.youtube.com/watch?v=669AcEBpdsY
[30] http://themindunleashed.org/2013/12/researchers-reveal-stonehenge-stones-hold-incredible-musical-properties.html
[31] https://www.cymascope.com/cyma_research/egyptology.html
[32] http://www.dancingwithwater.com/about-the-book/
[33] http://faculty.washington.edu/ghp/research-themes/water-science/
[34] http://faculty.washington.edu/ghp/cv/
[35] http://www.i-sis.org.uk/liquidCrystallineWater.php
[36] https://www.youtube.com/watch?v=nTngSAG28J4
[37] http://charleseisenstein.net/the-waters-of-heterodoxy-g-pollacks-the-fourth-phase-of-water/
[38] http://www.jeffereyjaxen.com/blog/the-4th-phase-of-water-a-key-to-understand-all-life
[39] https://www.youtube.com/watch?v=45yabrnryXk
[40] http://www.wikiwand.com/en/Double-slit_experiment
[41] http://www.simonsfoundation.org/quanta/20130917-a-jewel-at-the-heart-of-quantum-physics/
[42] http://resonance.is/
[43] https://youtu.be/orqJZCGdCh4
[44] http://www.wikiwand.com/en/Quantum_entanglement
[45] http://www.wikiwand.com/en/Quantum_tunnelling
[46] http://www.wikiwand.com/en/Solar_System
[47] http://www.universetoday.com/107322/is-the-solar-system-really-a-vortex/
[48] https://www.youtube.com/watch?v=zBlAGGzup48
[49] https://www.youtube.com/watch?v=4cgQNUhtmHM
[50] http://www.umass.edu/newsoffice/article/wang-international-team-discover-why-super
[51] http://resonance.is/stephen-hawking-goes-grey/

[52] http://www.wikiwand.com/en/Gravity
[53] http://www.wikiwand.com/en/Proxima_Centauri
[54] http://www.wikiwand.com/en/Magnetic_reconnection
[55] http://www.wikiwand.com/en/Endangered_species
[56] http://www.wikiwand.com/en/Elephant#/Threats
[57] http://wwf.panda.org/what_we_do/endangered_species/elephants/african_elephants/
[58] http://www.nytimes.com/2012/09/04/world/africa/africas-elephants-are-being-slaughtered-in-poaching-frenzy.html?_r=4&ref=world&
[59] http://www.nonhumanrightsproject.org/
[60] http://news.nationalgeographic.com/news/2006/10/061030-asian-elephants.html
[61] http://www.wikiwand.com/en/Brain_size
[62] http://www.wikiwand.com/en/Elephant
[63] http://www.birds.cornell.edu/brp/elephant/about/about.html
[64] http://www.wikiwand.com/en/Animal_language
[65] http://www.wikiwand.com/en/Elephant_cognition
[66] http://www.wikiwand.com/en/Ivory_trade
[67] http://www.wikiwand.com/en/Lost_Decade_%28Japan%29
[68] http://www.wikiwand.com/en/Japanese_asset_price_bubble
[69] http://ens-newswire.com/2014/02/12/obama-bans-u-s-commercial-trade-of-elephant-ivory/
[70] http://www.theguardian.com/environment/2014/apr/25/kenya-drones-national-parks-poaching
[71] http://www.nonhumanrightsproject.org/
[72] http://www.wikiwand.com/en/Apex_predator
[73] http://www.sharksavers.org/en/education/biology/450-million-years-of-sharks1/
[74] http://www.wikiwand.com/en/Shark
[75] http://www.wikiwand.com/en/Sharks_in_captivity
[76] http://www.sharkbookings.com/why-great-white-sharks-cannot-be-kept-in-captivity/
[77] http://www.wikiwand.com/en/Great_white_shark
[78] http://www.wikiwand.com/en/Ampullae_of_Lorenzini
[79] http://animals.howstuffworks.com/fish/sharks/electroreception.htm
[80] http://greatwhiteshark3d.com/
[81] http://www.wikiwand.com/en/Shark_finning
[82] http://www.wildaid.org/
[83] http://www.wikiwand.com/en/Solon
[84] http://www.wikiwand.com/en/Seisachtheia
[85] http://www.wikiwand.com/en/Seisachtheia
[86] http://scholar.princeton.edu/sites/default/files/mgilens/files/gilens_and_page_2014_-testing_theories_of_american_politics.doc.pdf
[87] http://www.wikiwand.com/en/Themistocles
[88] http://www.wikiwand.com/en/Second_Persian_invasion_of_Greece
[89] http://www.wikiwand.com/en/Delian_League
[90] http://www.wikiwand.com/en/Athenian_democracy - /citenote7
[91] http://www.wikiwand.com/en/Socratic_method
[92] https://www.youtube.com/watch?v=vTyLSr_VCcg
[93] http://www.wikiwand.com/en/Electroencephalography
[94] http://electronics.howstuffworks.com/emotiv-epoc.htm
[95] http://www.heartmath.org/research/science-of-the-heart/head-heart-interactions.html?submenuheader=3
[96] http://www.heartmath.com/
[97] https://www.youtube.com/user/HeartMathInstitute/videos
[98] https://www.youtube.com/watch?v=O6FqZiLxhq4
[99] http://www.sciencedaily.com/releases/2004/08/040817082213.htm
[100] http://www.innerworldsmovie.com/index.cfm
[101] https://www.youtube.com/watch?v=qmL4CeTENtw
[102] Honey Goode, Research Assistant
[103] Honey Goode, Research Assistant
[104] http://www.innerworldsmovie.com/
[105] https://www.youtube.com/user/AwakenTheWorldFilm
[106] http://www.awakentheworld.com

[107] http://www.wikiwand.com/en/Epigenetics
[108] http://www.i-sis.org.uk/rnbwwrm.php
[109] http://www.i-sis.org.uk/Living_Rainbow_H2O.php
[110] http://www.i-sis.org.uk/Mystery_of_Missing_Heritability_Solved.php
[111] http://www.i-sis.org.uk/fromGenomicsToEpigenomics.php
[112] http://www.i-sis.org.uk/epigeneticInheritance.php
[113] http://www.i-sis.org.uk/epigeneticToxicology.php
[114] http://www.i-sis.org.uk/caringMothersGeneticDeterminism.php
[115] http://www.i-sis.org.uk/Living_Green_and_Circular.php
[116] http://www.i-sis.org.uk/Evolution_by_Natural_Genetic_Engineering.php
[117] http://www.i-sis.org.uk/Water_Structured_in_the_Golden_Ratio.php
[118] http://www.i-sis.org.uk/TIOCW.php
[119] http://www.i-sis.org.uk/DNA_sequence_reconstituted_from_Water_Memory.php
[120] http://www.i-sis.org.uk/liquidCrystallineWater.php
[121] http://www.i-sis.org.uk/Retracting_Serallini_study_violates_science_and_ethics.php
[122] http://www.i-sis.org.uk/Scientists_Declare_No_Consensus_on_GMO_Safety.php
[123] http://www.i-sis.org.uk/Global_Status_of_GM_and_nonGMs.php
[124] http://www.i-sis.org.uk/Why_GMOs_Can_Never_be_Safe.php
[125] http://www.i-sis.org.uk/Evolution_by_Natural_Genetic_Engineering.php
[126] http://www.i-sis.org.uk/Horizontal_Transfer_of_GM_DNA_Widespread.php
[127] http://www.i-sis.org.uk/How_mind_changes_genes_through_meditation.php
[128] http://www.i-sis.org.uk/No_Genes_for_Intelligence_in_the_Human_Genome.php
[129] http://www.i-sis.org.uk/How_to_Improve_the_Brain_Power_and_Health_of_a_Nation.php
[130] http://www.i-sis.org.uk/rnbwwrm.php
[131] http://www.i-sis.org.uk/Living_Rainbow_H2O.php
[132] http://www.theinstituteforventurescience.org/
[133] http://www.ebnerandsons.com/collections/books
[134] http://faculty.washington.edu/ghp/publications/
[135] http://faculty.washington.edu/ghp/research-themes/water-science/
[136] http://www.theinstituteforventurescience.org/
[137] http://www.ebnerandsons.com/collections/books
[138] http://www.theinstituteforventurescience.org
[139] http://www.epa.gov/climatechange/science/indicators/index.html
[140] http://www.antarctica.ac.uk/bas_research/science_briefings/icecorebriefing.php
[141] http://cdiac.ornl.gov/trends/co2/ice_core_co2.html
[142] http://www.wikiwand.com/en/Global_warming
[143] http://www.wikiwand.com/en/Scientific_opinion_on_climate_change
[144] http://www.amazon.com/Climate-Engineering-Boston-Review-Books/dp/0262019825
[145] http://www.wikiwand.com/en/Stratospheric_sulfate_aerosols_%28geoengineering%29
[146] http://www.wikiwand.com/en/Solar_radiation_management
[147] http://www.skyderalert.com/
[148] http://www.i-sis.org.uk/Unintended_Hazards_of_Geoengineering.php
[149] http://www.i-sis.org.uk/GeoEngineeringAMD.php
[150] http://www.scidev.net/global/earth-science/news/concerns-grow-over-effects-of-solar-geoengineering.html
[151] http://phys.org/news/2013-03-responsible-geoengineering-experts.html
[152] https://www.youtube.com/watch?v=jHm0XhtDyZA